From the Lab into the World

By the Same Author

SCIENTIFIC MONOGRAPHS

Optical Rotatory Dispersion: Applications to Organic Chemistry
Steroid Reactions: An Outline for Organic Chemists (editor)
Interpretation of Mass Spectra of Organic Compounds
(with H. Budzikiewicz and D. H. Williams)
Structure Elucidation of Natural Products by Mass Spectrometry
(2 volumes, with H. Budzikiewicz and D. H. Williams)
Mass Spectrometry of Organic Compounds
(with H. Budzikiewicz and D. H. Williams)

NONFICTION

The Politics of Contraception
Steroids Made It Possible
The Pill, Pygmy Chimps, and Degas' Horse
From the Lab into the World: A Pill for People, Pets, and Bugs

POETRY

The Clock Runs Backward

FICTION

The Futurist and Other Stories
Cantor's Dilemma
The Bourbaki Gambit
Marx, verschieden (in German translation)

CREATORS OF MODERN CHEMISTRY

From the Lab into the World

A Pill for People, Pets, and Bugs

Carl Djerassi

American Chemical Society
Washington, DC 1994

Library of Congress Cataloging-in-Publication Data

From the lab into the world: a pill for people, pets, and bugs / Carl Djerassi, author

p. cm.

Includes bibliographical references and indexes.

ISBN 0–8412–2808–6

1. Contraceptive drugs—Research.

I. Title.

RG137.4.D54 1994
615'.766'0724—dc20

94–17603
CIP

The paper used in this publication meets the minimum requirements of American National Standard for Information Sciences—Permanence of Paper for Printed Library Materials, ANSI Z39.48–1984. ∞

PRINTED IN THE UNITED STATES OF AMERICA

for my grandson
Alexander Djerassi

Series Advisory Board

Contents

SCIENTIFIC COOPERATION AND THE DEVELOPING WORLD

MISCELLANEOUS TOPICS

About the Author

Carl Djerassi is a chemistry professor at Stanford University, although he has retired as an active research chemist to pursue numerous other interests: research on public policy issues; writing fiction; and active involvement in the Djerassi Resident Artists Program, an artists' colony located south of San Francisco.

Among many national and international honors, Djerassi received the Priestley Medal, ACS's highest award, for his work in many areas of chemistry, including development of new medicinal compounds (such as antihistamines and topical corticosteroids), research on hundreds of natural products, and development of methods (such as optical rotatory dispersion, circular dichroism, mass spectrometry, and computer artificial intelligence techniques) for solving stereochemical and structural problems. He is probably best known for the synthesis and subsequent industrial development of the first oral contraceptive at Syntex in 1951, for which he was awarded the National Medal of Science. His many research activities have led to publication of more than 1200 papers and 13 books, of which he has been author or coauthor, including an autobiography, *The Pill, Pygmy Chimps, and Degas' Horse.*

Preface

Tricky business, assembling in book form a collection of essays written over the course of a quarter of a century, especially when the author and organizer are the same person. Before descending to explanations—usually only one step removed from excuses—I shall start at the elevated level of personal pleasure and pride. When the American Chemical Society approached me with the idea of publishing a collection of my essays, I was taken aback. Until that moment, I had not realized that I had written almost as many essays as I bore years—the proverbial three score and ten—nor had I ever assembled them or reread them as a group. Now having done so, I am pleased with myself for having taken the time, during an intense working life focused on producing hundreds of highly technical scientific articles, to write on socially relevant topics with a general audience rather than peer specialists in mind. "Not bad," I thought, "there is a lot of material a general readership might find instructive."

But it is easier to fill a vessel, I soon realized, than getting someone to drink or even sip from it. Organizing such supposed distillates of wisdom or opinionated messages—depending on one's characterization of my writing—is, as I said, tricky. Should the assemblage be chronological or topical? Should the individual pieces be left undisturbed, thus truly reflective of the time and the then existing authorial persona, or revised in terms of today's circumstances? And finally, in spite of my present autobiographical frame of mind, reflected in two published autobiographies and several quasi-autobiographical novels, is it really worthwhile to relaunch these essays in one package?

My answer to the last question is an unequivocal "yes," because the major portion deals with an overwhelmingly important theme: birth control (of humans; of insects; even of pets)—a subject with which I have been intimately associated over the decades as a research scientist, as an industrial entrepreneur, as a teacher, and as a constructive polemicist. Practical answers to these problems are so complicated, and their application so interwoven with political and societal threads, that thoughtful essays, some written as long ago as the 1970s, are still, in most instances, relevant to the 1990s. It is—or should be—a humbling experience for author and reader alike to realize how little has changed in terms of fundamentals during those years. (If the problems of human birth control are too heavy or troubling a starting point, I suggest that more sensitive readers peruse Chapter 18, "Planned Parenthood for Pets?", to see how little has changed over the course of two decades).

This brings me to the tricky business of reading such a volume: it is essential to do so with chronologically bifocal glasses. Little is gained by scrutinizing analyses and prognoses written in the late 1960s or early 1970s through glasses polished in the 1990s. A crucial message, however, comes across by examining the problems of the 1990s through 1970 spectacles: the most important problem of contemporary life—the growth of the world's population—is an intrinsically gray problem, not amenable to black and white answers. In retrospect, no one—academic scientists, media, public, and least of all, politicians—was or is prepared to face the true cost of time. If there is one leitmotif to the present volume, it is my emphasis on this lesson.

In the end, partly owing to the highly efficient and professional assistance provided by Janet S. Dodd (senior editor of ACS Books) and Elizabeth Wood (ACS copy editor), the organization of this volume fell into place with remarkably little trauma or delay. The overall arrangement is chronological within certain logical subtopics, because I wanted to transmit a sense of how my own attitudes and assumptions changed over time. Some essays were deleted because of redundancy; others were combined for the sake of conciseness, and a few were revised for the sake of historical cross-reference.

Of the three main topical divisions—birth control and contraception; bridging the technological gap between the haves and have-nots; and some miscellaneous topics—the first two require no further commentary. For decades of my professional life as a research scientist, they were the focus of my social conscience. The miscellaneous category, however, does warrant some brief comment, because it addresses topics and activities that have increasingly occupied me during the past decade.

For instance, the role of women in contemporary society, and especially in science, has greatly affected my teaching at Stanford University (moving from Chemistry to the programs in Human Biology and in Feminist Studies) as well as the fiction writing that has become my latest and perhaps most intense intellectual preoccupation. Two other topics that have become a focus of my lecturing and, above all, my fiction, have been the manner in which research scientists transmit their behavioral characteristics through the mentoring process, and the manner in which we indulge in competing interests or eschew them. Though I am surely in the minority in believing that professional bigamy—the source of most conflicting interests—can also be a virtue and not solely a vice, it has not kept me from expressing that view forcefully in some essays (one of which is reprinted in this volume) and in my autobiographical writing. I believe that intellectually speaking, a polygamous range of interest is even more commendable, which is one reason why I have ended this collection with an essay on Art patronage, the last portion of which is deeply personal.

I conclude with a pithy *caveat lector* embodied in a prophetic fortune cookie message, applicable to much of this book's subject matter. I encountered it in a San Francisco Chinatown restaurant and first alluded to it in my 1975 Perkin Medal Award address. But since the latter is not included in this collection, the Confucian warning bears repeating: "Your problems are too complicated for fortune cookies."

Carl Djerassi
June 1994

1 From the Lab into the World

In spite of pronounced tendencies toward chemical workaholism, I was never an inhabitant of a scientific ivory tower uncontaminated by events from the outside world. My immigration in 1939 to the United States as a teenage refugee from Hitler's Europe had permanently sensitized me to outside political events. The fact that my research career started in the pharmaceutical industry—one year at Ciba in Summit, New Jersey, right after graduation from Kenyon College and another four years with the same company immediately after my Ph.D. from the University of Wisconsin—had given me an understanding of the practical applications of laboratory research, an awareness that stood me in good stead when I started on an academic career and eventually led me to become a professional bigamist by occupying concurrently important positions in industry and academia.

A seminal event during those early post-Ph.D. years was my decision in late 1949 at age 26 to accept a position as associate director of chemical research at Syntex—at that time a minute chemical manufacturer of steroids in Mexico City. Within two years, we published the first synthesis of cortisone from a plant raw material, announced the first synthesis of an orally active progestational steroid that became the key component of contraceptives that are used to this day by millions of women, and published several dozen papers on rather complicated ste-

Note: Original text written in 1992.

roid projects—all of this from the chemical wilderness of Mexico. Even though I then returned to the United States to start a professorial career at Wayne State University, followed (after another three-year stint at Syntex in Mexico from 1957 to 1960) by my move to Stanford University, my years in Mexico left an indelible imprint on me in terms of recognizing the gulf between the scientific "haves" and "have-nots."

Already in the 1950s, largely through assistance from the Rockefeller Foundation, I directed collaborative projects with groups at the National University in Mexico City, and shortly thereafter at the Instituto de Química Agricola in Rio de Janeiro that featured residencies of one or more years by experienced postdoctorate fellows from my laboratory in those Latin American institutions, coupled with periodic short-term visits on my part. The purpose of these collaborative ventures was to demonstrate to young Mexican or Brazilian chemists that significant research could be conducted in their own countries rather than solely under the comparatively luxurious circumstances existing in high-class American universities. These joint research enterprises worked so well that during the 1960s they served as prototypes for two projects of much wider breadth, which I helped initiate through the intermediacy of two organizations with which I had become associated during that decade.

The first was the Latin America Science Board of the National Academy of Sciences (NAS), which, as the name implied, sponsored scientific interactions with Latin American countries. Harrison Brown, a distinguished geochemist from the California Institute of Technology and perhaps the most activist foreign secretary the NAS ever had, invited me to join that board and eventually to serve as its chairman.

A few years later, still in the 1960s, it was folded into BOSTID—the academy's Board on Science and Technology for International Development—an Agency for International Development (AID)-funded program that encompassed virtually all bilateral scientific cooperative programs of the National Research Council with less developed countries in Latin America, Asia, and Africa. It was one of the few prestigious scientific organizations that departed from the conventional East–West travel axis of scientists to concentrate on North–South interactions—an emphasis that fit in very well with my Mexican and Brazilian experiences and that, at the same time, allowed me to expand my scientific geographic horizon to countries such as Colombia, Zaire, Sri Lanka, and Indonesia—part of the time as chairman of BOSTID.

Under Brown's tenure as foreign secretary of the NAS, the Brazilian Research Council undertook a number of science policy conferences with our group. At one such meeting, it outlined its high-priority goal of upgrading graduate education in various chemical subdisciplines because of the anticipated ripple effect on economically significant fields in industry, agriculture, and health. This prompted me to propose an extension of our Stanford–Rio project, which had focused on the chemistry of alkaloids and other natural products from Amazonian plants, to those other chemical areas of concern to the Brazilians. Why not invite a group of American university professors to serve as part-time research directors in the same manner as I had functioned in the past, I asked, and have them send to Brazil postdoctorate fellows from their laboratories as in situ tutors of local graduate students?

In retrospect, it is amazing how fast we managed to cut through bureaucratic and fiscal knots and how quickly this United States–Brazil program became operative with the enthusiastic participation of a number of American chemical superstars, such as William S. Johnson, John Baldeschwieler, Aaron Kupperman, Charles Overberger, and three Priestley Medalists (George Hammond, Harry Gray, and Henry Taube)—to name just some of the American part-time research directors from Stanford, Caltech, and the universities of Indiana and Michigan, who, with the exception of Kupperman, had never been to Brazil. They provided the overall guidance in synthetic organic, polymer, inorganic, and experimental physical chemistry and sent postdoctorate fellows to the laboratories of their Brazilian counterparts to help in the training of graduate students and in the conduct of advanced research in their specialties.

After a few years, I turned over the chairmanship of this program to the distinguished Brazilian–American physical chemist, Aaron Kupperman of Caltech, who continued in that position until the maturation of this project under which a substantial number of Brazilians earned M.S. and Ph.D. degrees in São Paulo and Rio de Janeiro, the two university sites where the program operated. Twenty years after its initiation, the chairmen of the chemistry departments in both Rio (Bruce Kover) and São Paulo (José Riveros) were products of that binational chemical collaboration.

The extraordinarily capable staff director of BOSTID was Victor Rabinowitch, who, together with Harrison Brown, had been an active American member of the Pugwash Movement. It was through their initiative that I was invited to the 1967 Pugwash Conference on Science and World Affairs in Ronneby, Sweden, which offered me a second broad forum for fostering North–South collaboration. Pugwash was the outcome of the famous 1955 Russell–Einstein manifesto, which started as follows: "In the tragic situation which confronts humanity, we feel that scientists should assemble in conference to appraise the perils that have arisen as a result of the development of weapons of mass destruction, and to discuss a resolution in the spirit of the appended draft."

As expected, the first few conferences (named after the Nova Scotian village of Pugwash, where the first meeting was held) involved primarily physicists, because the main topic was control of nuclear weapons; but among the early participants were also some chemists such as Linus Pauling, another Priestley Medalist; Harrison Brown; Eugene Rabinowitch (Victor's father); Paul Doty; and Frank Long. By the mid-1960s the scope of the Pugwash conferences had widened from disarmament issues to include the increasing gap between rich and poor countries.

This is why I was invited by the American Pugwash Committee, which was then dominated by scientists from the Northeast (and especially Boston), to attend the Ronneby meeting. Pugwash participants are encouraged to prepare discussion papers, which are distributed ahead of time to expedite the flow of ideas. My own professional competence was outside the area of arms control, but I had ample experience—from my life in Mexico, from my research in Brazil, and from my travels—in the ever-widening North–South gulf, especially as it pertained to science and technology. I took the Pugwash invitation seriously; I felt the time was ripe to generalize before an international audience from my personal experiences of the conduct of scientific research in Mexico and Brazil.

I did so in a paper entitled "A High Priority? Research Centers in Developing Nations" (subsequently published in the *Bulletin of the Atomic Scientists*). I suggested that, from the standpoint of scientific development, a "developing" country becomes a "developed" one when original research emanates from it. The ability to perform such advanced research, be it in a university or other research center, usually appears at the end of the list of priorities for developing countries. In light of the ever-increasing tempo of scientific and technological progress in the advanced countries, the creation of competitive, basic research centers in a developing country through the traditional routes becomes a hopeless proposition. This is particularly true in scientific research, where there exists only one standard of excellence. Saying "this is very good chemical research for Kenya, but rather poor for Sweden" is like saying that poor chemical research is being performed in Kenya.

It is ironic that I should have arbitrarily chosen Kenya as an example of a developing country then lacking the requisite indigenous scientific manpower, to make the following proposal: "(1) select an international cadre of postdoctorate research fellows; (2) provide overall scientific direction by a group of part-time directors from major universities in different developed countries; and (3) select research areas with a possible ultimate economic payoff and a maximum multiplication factor."

Surprisingly, my proposal was taken up not by a scientist from one of the affluent countries, but rather by an African entomologist, Thomas Odhiambo of Nairobi University. He wrote to me in 1968: "Can a move be made to develop one such center of excellence in mid-Africa, for example, in Nairobi? At the risk of appearing presumptuous, I would like to see such a center—on insect physiology and endocrinology—established in Nairobi. . . . Can you suggest how to achieve this? Would you be prepared to help launch such a scheme?"

Even under ordinary circumstances I would probably have found it difficult to refuse such a challenge. But 1968 was the year of the insect in my personal Chinese calendar, when I was about to lead—as chairman of the board and chief executive officer—the newly founded Zoecon Corporation's exploration for applications of recent advances in insect endocrinology. In addition, a few years earlier, I had fallen in love with East Africa during two trips with my children, when we roamed the game parks in Uganda, Tanzania, and Kenya.

At the suggestion of Victor Rabinowitch, I contacted the American Academy of Arts and Sciences in Boston, then also the home of the American Pugwash Committee. Its general secretary, John Voss, managed to convince the governing council of his academy to fund Odhiambo's travel to Boston. The purpose of the trip was for Odhiambo to meet a few American insect biologists—notably Harvard's Carroll Williams, who was then a consultant for Syntex—and several other scientists from the Boston area and from Cornell.

Key to the success of my Ronneby proposal was the willingness of scientists from advanced countries to serve as part-time research directors—as I had done in Mexico and Brazil. Only then would it be possible to attract young postdoctorate fellows to some new research center thousands of miles from their home bases, both to help train local scientists and to establish quickly an institution of high visibility.

Carroll Williams, one of the pioneers in insect hormones, was a prime candidate for such leadership, and his participation was likely to sway other "prima donnas" of the insect world. Odhiambo, charismatic and intelligent, immediately seduced Williams and several colleagues into examining the feasibility of my proposal in situ, whereupon Voss and Rabinowitch went into high gear to organize a meeting in Nairobi.

We raised some money from various philanthropic sources to cover the travel expenses of a number of Americans. Once we knew that American involvement was ensured, Voss and Rabinowitch contacted the officers of several foreign academies—among them the Royal Society of London, the Dutch Academy, and most important (as it turned out), the Royal Swedish Academy of Sciences—and persuaded them to join. Out of this planning meeting in Nairobi in the fall of 1969 came ICIPE—the International Center for Insect Physiology and Ecology, a remarkable example of international cooperation by a consortium of 21 national scientific academies.

In the intervening decades, ICIPE has become an internationally known center of insect science, still headed by Tom Odhiambo, who now deals with multimillion-dollar budgets. My original prescription for the genesis of an instantaneous oasis in a scientific desert could have turned into benevolent do-goodism, or even worse, into some sort of scientific neocolonialism. But it did not work out that way; both Odhiambo and ICIPE's governing council understood that the Africanization of the enterprise had to be the ultimate criterion of its success, and it is now largely staffed and directed by African scientists.

The ICIPE story finally brings me to 1969, which I categorize as a watershed year in the development of my social conscience. During the 1950s and 1960s, I had become increasingly involved in thinking and doing something about the ever-increasing gap between rich and poor countries. But all of my efforts, and most of my thoughts, focused on scientific and technological aspects of that problem—in other words, on sociogeographical extensions of my hard chemical research.

Thus far I have barely touched on that "hard" chemical research, which occupied the major portion of my intellectual life, and I do not propose to discuss it now. In the mid-1980s, Jeffrey Seeman, editor of the American Chemical Society series of chemical autobiographies entitled Profiles, Pathways, and Dreams, planted an autobiographical virus in me, which metastasized rapidly just after I started to recover from the most serious illness of my life. The first concrete manifestation of that viral infection was my own contribution to Seeman's ACS-published book series, which appeared in 1990 under the title *Steroids Made It Possible.* That book is a typical autobiography of a scientist, addressed to other scientists, and replete with excruciating details of his scientific contributions that are clothed in a very thin garment of modesty and embellished with many chemical structures, yet it discloses relatively little of the author's private life and beliefs.

In my own case, this lack of nonscientific particulars can be attributed to a much wider spread of Seeman's virus in that it prompted me to publish in 1992 a different type of autobiography, addressed to a general, nontechnical audience, under the title *The Pill, Pygmy Chimps, and Degas' Horse.* Some of the chapter headings, such as "Freud and I," reveal that this second autobiography covers much broader and more intimate terrain than the usual scientific autobiography.

The year 1969 was the climax of my professional polygamy: I had assumed the position of president of Syntex Research, a few years after the company had moved its administrative and research headquarters to the Stanford Industrial Park, a few minutes from my Stanford University office and laboratory. Officially, I became a half-time professor, but in actual fact I reduced neither my academic research nor my teaching duties.

In addition, I helped found and then headed Zoecon Corporation, also located on the Stanford Industrial Park, and served as chairman of a third company I had helped spawn, Synvar—a joint partnership between Syntex and Varian—that eventually changed its name to Syva and became a well-known, innovative diagnostic company.

In the late 1960s, a significant portion of Syntex's research expenditures was still dedicated to what, during that decade, could easily be considered globally the most pressing area of medicine: human fertility control. In my new position as president of Syntex Research, I was concerned not only with basic research, or with the management of what was then the largest industrial postdoctorate fellowship program in any pharmaceutical company, but also with many applied areas, including toxicology, clinical research, and interaction with the Food and Drug Administration (FDA). In addition, Syntex-developed oral contraceptives had been licensed to Ortho, Eli Lilly, Parke-Davis, and Schering, so that I was aware of attitudes toward contraceptive R&D in other pharmaceutical companies as well as within the World Health Organization (WHO) through occasional participation in its Special Programme of Research, Development and Research Training in Human Reproduction.

The first storm clouds with respect to society's attitude toward contraceptive research had started to appear around that time, which also saw the flowering of three important social movements concerned with women's rights, environmental concerns, and consumer advocacy. All three displayed a psychologically understandable suspicion of technology and, indirectly, also of science; and all three achieved many of their aims largely through the unique character of the U.S. litigation system.

When around that time the first substantial epidemiological studies raised questions about the Pill's less obvious side effects, women, who earlier had objected to being used as human guinea pigs, now asked why the Pill had not been tested more thoroughly—a question that was also raised in the press and in legislative commentary. The interplay among the public, the press, legislators, and the FDA is full of feedback mechanisms, each of which stimulates and restricts the other.

When it came to contraceptive R&D in the late 1960s, there were few stimulations. The restrictions predominated, largely because the lay constituencies—the public, the media, and the legislators—demanded black-and-white answers to intrinsically gray questions of enormous complexity. I can think of no field where personal risk–benefit considerations are as overwhelming as in matters dealing with sex and contraception.

I picked a peculiar venue, the 1969 Pugwash Conference in Sochi (USSR) on the Black Sea, to warn about the impact of these public concerns on the future of contraception. I incorporated that talk into my first public policy paper on contraception, which appeared that year in *Science* under the title "Prognosis for the Development of New Chemical Birth Control Agents."

A year later, just after the conclusion of the notorious "Nelson hearings" in the U.S. Senate on the safety of contraceptives, which were sensationalized by the press (thus giving the contraception field an extremely poor image), Harrison Brown organized a symposium at Caltech on "Technological Change and Population Growth." I used that forum to launch what I still consider my most influential contribution to public policy: "Birth Control after 1984," which also appeared in *Science*. No other paper of mine, chemical or nontechnical, received as many reprint requests or was as widely reprinted in other books as this one. It appears herein as Chapter 5.

As I stated in the first paragraph, "It behooves us to consider what some of the future contraceptive methods might be and especially what it might take, in terms of time and money, to convert them into reality. There are many publications on this subject, but none seems to have concerned itself with the logistic problems associated with the development of a new contraceptive agent." I then proceeded with a detailed, logistic prognosis for the development of two types of fundamentally new, yet scientifically feasible, contraceptive agents: a female once-a-month pill with abortifacient or menses-inducing properties, and a male pill.

I had chosen the date "1984" in my title not only for its Orwellian overtones, but because I wanted to emphasize that it would take an average of 14 years from initiation of laboratory research for such a new contraceptive to final FDA approval for wide public use. (It is interesting that the French-developed RU-486—an approximation of the female pill I hypothesized back in 1970—was introduced into medicinal practice just about 14 years later!) I ended the article with a set of recommendations, without which the continued participation of the pharmaceutical industry in this field would largely disappear, and concluded with the words, "Birth control in 1984 will not differ significantly from that of today [1970]."

A couple of years later, shortly before receiving the National Medal of Science from President Nixon for our first synthesis of a steroid oral contraceptive, I announced in my capacity as president of Syntex Research that it did not make commercial sense for Syntex to continue to spend money on R&D in the contraceptive field. In the late 1960s, 13 large pharmaceutical companies (9 of them American) had meaningful research commitments in the field of birth control; by the mid-1980s only four were left (one of them American).

I had become convinced that politics, rather than science, would play the dominant role in shaping the future of that field—a personal conclusion that not only affected my decisions as a research director in industry, but also caused a dramatic shift in my role as a university teacher. Until then, I had only taught a variety of chemistry courses, primarily advanced ones directed at graduate students. But if politics starts to have a negative impact on a technical field, and especially on one of high societal importance, then it seemed to me that the most constructive action I could take was to educate the decision makers and politicians of the future, to which Stanford University invariably contributes a significant number.

In 1972 I volunteered to offer a course entitled "Biosocial Aspects of Birth Control" for advanced undergraduates in our Human Biology Program—a then newly instituted, interdisciplinary program that within a few years became one of the most popular undergraduate majors at Stanford. I was, and still am, the only chemistry professor to join its faculty, which otherwise bridges the "hard" and "softer" sciences by having representatives from biology, psychology, sociology, and anthropology, as well as many medical school departments. My participation in this undergraduate program eventually led to a total change in my life as a classroom teacher, an experience that I describe in detail in my autobiography under the heading "Condoms for the Teacher."

Of my several aims, the most important was to encourage students to think seriously about public policy in the context of real problems. My only requirement was that students be seniors and thus competent in at least one discipline (e.g., biology, religion, economics, political science) relevant to my chosen topic: human birth control.

I had a special educational experiment in mind, for which I needed students with adequate representation from various ethnic, social, and religious backgrounds and equal distribution by gender. There would be no examinations, I announced, and my formal lectures (three hours at a time) would end after two weeks. The rest of the time would be spent in a research mode in task forces consisting of up to six students each, who would focus on projected improvements in birth control of very specific population groups, because I wanted these future citizens and opinion leaders to realize that the concept of an ideal, universal birth control agent or approach was a chimera.

Because of the tremendous divergence of different populations, what is appropriate for one group or even one individual may not suit the next. What the world needs, figuratively and even literally, is a contraceptive supermarket; and I wanted the students to propose, through their own research, what some of the components of that supermarket might be.

Each student task force thus included majors from the various social sciences as well as premedical students. Although all important social and technical advances in real life are the result of interdisciplinary team efforts, we tend not to incorporate that concept in our undergraduate curricula. Our evaluation system emphasizes individual performance and competition; collaboration among students is explicitly or implicitly considered cheating.

In my course, however, I insisted that students collaborate: They were expected to organize their research together, with each task force member then writing a separate chapter of the group's report from her or his professional perspective. During the research phase, I met twice weekly with each student group, at which time I questioned them about research progress, provided them with key contacts or references, and encouraged them with funds from a modest financial kitty to use long-distance telephone calls as the most rapid way of extracting information from government bureaucrats here and abroad.

The climax of the course was the presentation of each task force's conclusion to the rest of the class and some invited guests over a period of three hours—half for formal presentations, the other for questions and answers. The range of task force topics was extraordinarily wide: Many groups selected specific population subgroups such as white American college students (typified by the majority of Stanford's affluent student population), Chicanos in San Jose, Chinese in San Francisco's Chinatown, or Puerto Ricans in Manhattan. Others chose functional rather than geographic or ethnic divisions: birth control problems of carriers of genetic diseases or of developmentally disabled persons. The manner in which the research results were presented in class was almost as diverse as the topics: Students used a variety of audiovisual aids, which frequently were combined with skits or even original short plays.

The most ambitious projects were conducted by my third class, in 1976. Stanford has a fairly sophisticated student evaluation of course contents and teacher performance. By the mid-1970s I had received feedback from two earlier classes of students who claimed that "Biosocial Aspects of Birth Control" as a one-quarter course had demanded more work than any other class in their undergraduate experience. I had found the same to be true of my professorial experience: Because each group worked on a different project, I had to be prepared to cover an extraordinarily wide range of subjects and to read, criticize, and grade the final reports from each task force, which were usually at least 100 pages long with many dozens of references.

"Why not spread the course over two quarters?" the students asked. "Why not?" I replied, and promptly turned to the Rockefeller Foundation for funds to cover the travel expenses of my human biology class for some exploratory research in more distant locations. I was thus able to organize the largest class of all, consisting of 10 task forces, and to offer each group the opportunity of having two or more members travel to sites irrespective of distance.

Geographically, the most ambitious task forces chose populations in Java, Kenya, and Mexico, but some of the U.S.-based task forces studying Indians in New Mexico or a rural population in the Deep South also came up with intriguing projects. The quality of some of the task force reports was so high that one of the students ended up in a job with the AID in Washington, another on the staff of the NAS, and a third with the WHO in Geneva.

Eventually, I extended this teaching approach to two other courses. Pesticides are clearly one of the causes of the general public's chemophobia, and because I had learned a lot about the challenges of developing novel approaches to insect control through my industrial activities with Zoecon, I offered a similar course on "Pest Control: Technical and Policy Aspects," in which the task forces had to focus on separate California crops, say cotton or rice, instead of human population groups. I still recall the memorable task force presentation conducted one evening at a premium California winery by a group of students studying insect control problems in vineyards.

The other extension of this teaching approach has led me to create a course on "Feminist Perspectives on Birth Control" in Stanford's Feminist Studies Program, which I have taught several times. In the mid-1980s one of the task forces in that course was responsible for the introduction of condom-dispensing machines at Stanford.

These teaching activities and the resulting interaction with a much broader range of students than is usually encountered in chemistry classrooms were in part responsible for my ever-increasing interest in the "softer" aspects of the technoscientific areas that had preoccupied me over the course of decades as a "hard" scientist. This gradual change caused me to focus increasingly on the cultural and behavioral aspects of a research scientist's life—what I once called the "soul and baggage of contemporary science." In an attempt to explain them to the scientifically illiterate, I first had to be sure that I could define these features to myself. This personal discourse became a public one as I decided to illustrate the scientist's tribal culture through the medium of an infrequently used literary genre that I prefer to categorize as "science-in-fiction" to differentiate it from science fiction.

It is ironic, but also refreshing, that after nearly five decades in science—by definition a profession devoid of fiction and solely dependent on monologuist written discourse—I am starting a new intellectual life in which I use the dialogist style of the fiction writer to talk about the truth behind the scientific persona. I find this as challenging as many a laboratory discovery.

Birth Control and Contraceptive Research

2

Parentage of the Pill

Our phallocentric society invariably focuses on the patrimony of scientific discovery, of a new drug, of the Pill . . ., searching for the "Father of . . .". But the birth of a drug, first and foremost, requires a mother, and most of the time also a midwife or obstetrician. Every synthetic drug, including steroid oral contraceptives, must start with an organic chemist. Until she or he has invented it, that is, conceived its chemical structure and then synthesized the molecule, nothing can happen. This is the reason why I maintain that in the parentage of any synthetic drug, including the ovulation-inhibiting progestational constituent of the Pill, the organic chemist—irrespective of gender—symbolizes the mother, with the chemical entity representing the egg.

Only then does the biologist enter the picture, performing a variety of biological experiments that I equate to sperm floating around the ovum. The key experiment, confirming the anticipated biological activity or demonstrating some unexpected new one, can then be considered the sperm associated with the actual fertilization. Thus, in my picture, the biologist—again regardless of gender—plays the paternal role, and the clinician's subsequent efforts correspond to obstetrical and pediatric functions in the development and maturation of a drug.

The development of the ovulation-inhibiting component of the Pill represents one of the few instances where the chemical contributions

Note: Original text written in 1978, 1984, 1992 and 1993.

did not result from chance or serendipity but rather were fairly predictive. The role of the natural female sex hormone, progesterone, was known since the early 1930s; it is nature's contraceptive—the reason that women do not conceive again during pregnancy is that progesterone is continuously secreted at that time—and various biological scientists had from time to time considered progesterone's ovulation-inhibiting properties as one possible approach to contraception. But one of the complications of progesterone is that it is essentially inactive orally: To use it therapeutically, whether for fertility control or other purposes, daily injection was necessary.

THE ROUTE TO AN ORAL PROGESTATIONAL COMPOUND: NORETHINDRONE

At the time that I became interested in the chemistry of progestational steroids, one of the tenets of steroid chemistry was that almost any chemical alteration of this molecule—in contrast to the situation with the estrogenic hormones—would either diminish or destroy biological activity. But in 1946 the chemist Maximilian Ehrenstein from the University of Pennsylvania reported the first synthesis of a mixture of 19-norprogesterones (a mixture of compounds that differed structurally from the natural hormone by the elimination of a methyl group), which was found to be active. This was a remarkable observation in terms of our previous assumptions about progesterone.

One day in the spring of 1949, I received an employment offer from Syntex, a company of whom I had never heard before. Although the position (as associate director of chemical research) seemed tempting to me (not yet 26, I had already five years of industrial experience at CIBA), the location of Syntex in the chemical desert of Mexico made the offer seem ludicrous. Fortunately, my touristic inclinations were already highly developed at that time and, because no prior commitments on my part were associated with the invitation, "Come and visit us in Mexico City with all expenses paid," I went.

George Rosenkranz, then technical director of Syntex and barely past 30, impressed me enormously as a sophisticated steroid chemist and also charmed me personally. Rosenkranz showed me rather crude laboratories (I was convinced the homemade hoods had no fans, depending solely on natural ventilation), but he promised lots of laboratory assistants and substantial research autonomy. Furthermore, even though the labs were primitive (I still recall my charmed amusement on observing a hydrogenation vessel shaking in the sunshine of the open patio on Calle Laguna Mayran 413), some of the Syntex equipment was quite advanced for its days, an example being the availability of a single-beam Perkin Elmer infrared spectrometer at a time when neither CIBA nor my alma mater, the University of Wisconsin, had such an instrument, which proved to be enormously useful for steroid research.

Part of my Ph.D. thesis at the University of Wisconsin in the early 1940s had dealt with the partial synthesis of the then inaccessible estrogenic hormones from the more readily available androgens. Hans H. Inhoffen, at Schering A.G. in Berlin, had demonstrated the practical feasibility of such a partial aromatization of ring A, but the work had been performed during the war and experimental details were scant. Syntex had started to use the Inhoffen process (which had not been patented in Mexico) for the production of modest quantities of estrone and estradiol. I suggested, and Rosenkranz concurred enthusiastically, that we examine another and potentially proprietary route to the estrogens directly from testosterone. In less than three months we succeeded in accomplishing this aim.

Interestingly, this was only one of our more notable early research triumphs; our partial aromatization studies also turned into the impetus that led us in a fairly straight path to the first synthesis of an oral contraceptive. My Syntex colleagues in Mexico City and I became quite interested in following up on Ehrenstein's lead of 1946 and, using various chemical methods developed as part of our estrogen synthesis, prepared for the first time in 1951 pure 19-norprogesterone which, when tested at a commercial contract laboratory in Madison, Wisconsin, was found to be highly active. In 1939 another purely accidental discovery was made in Germany where chemists under the leadership of Inhoffen found that if an acetylene grouping is added to the male sex hormone testosterone, its biological activity is changed markedly: For unknown reasons this compound has weak progestational activity, and, most importantly, it is active by mouth. Putting these two observations together, on October 15, 1951, we made the 19-nor analogue of Inhoffen's compound—that is, 19-nor-17α-ethynyltestosterone or, for short, "norethindrone"—which turned out to be the first oral contraceptive to be synthesized.

But initially, we were not looking for an oral contraceptive when we developed an oral progestational compound. Our research was undertaken because at that time progesterone was used for treatment of menstrual disorders, for certain conditions of infertility, and at a research level, for the treatment of cervical cancer in women by local administration of a high dose of the hormone. This treatment was extremely painful because it involved injecting a fairly concentrated oil solution of large amounts of progesterone into the cervix, and we felt that a more powerful progestational compound that would be active orally would be a better candidate. As it happened, the progesterone treatment of cervical cancer did not pan out, but this motivation led us to make what was at that time the most potent, orally effective progestational compound.

SYNTHESIS OF NORETHYNODREL

On August 31, 1953, well over a year after our first publication dealing with the synthesis of norethindrone, Frank Colton of G. D. Searle & Company filed a patent for the synthesis of the double-bond isomer of norethindrone. Mild treatment of Colton's isomer, named norethynodrel, with acid, or just human gastric juice, converts it to a large extent in the test tube or in the body into Syntex's norethindrone. Is synthesis of a patented compound in the stomach an infringement of a valid patent? I urged that we push this issue to a legal resolution, but Parke-Davis, our American licensee, did not concur.

Searle was selling a very important anti-motion sickness drug, Dramamine, which contained Parke-Davis's antihistamine Benadryl, and our norethindrone in 1957 (the year the U.S. Food and Drug Administration [FDA] approved norethindrone and norethynodrel for the treatment of menstrual disorders and for certain conditions of infertility) seemed "small potatoes" over which it was not worth fighting with a valued customer.

Once we had established the high oral progestational activity of norethindrone, we supplied the substance to several endocrinologists, including Gregory Pincus of the Worcester Foundation for Experimental Biology in Shrewsbury, Massachusetts, for more detailed biological scrutiny. Among the many steroids tested in 1953 by his group at the Worcester Foundation for ovulation inhibition, norethindrone and norethynodrel were the two most promising candidates. Pincus, who was a consultant for Searle, selected the Searle compound for further work, whereas Syntex, not having any biological laboratories or pharmaceutical marketing outlets at that time, licensed Parke-Davis to pursue the FDA registration and market the product in the United States. It was only after 1957, when both norethindrone and norethynodrel had entered the market as drugs for noncontraceptive, gynecological purposes, that the paths of the two companies diverged.

Although Syntex-sponsored contraceptive trials with norethindrone were actively conducted in Mexico City and Los Angeles, Parke-Davis suddenly chose not to pursue these results through the FDA approval process, because of possible religious backlash, and returned the contraceptive marketing license to Syntex. A favorable licensing and marketing agreement was then negotiated by Alejandro Zaffaroni, Syntex's Executive Vice President, with the Ortho Division of Johnson & Johnson, a company with a long-standing commitment to the birth-control field, but the subsequent need to repeat certain primate studies that Parke-Davis was unwilling to hand over to Ortho caused a delay of nearly two years before Syntex's norethindrone received FDA approval for contraceptive indications. Finally, in 1964, three companies—Ortho, Syntex, and Parke-Davis (having changed their mind after realizing that no Catholic-inspired boycott had developed)—were marketing 2.0-milligram doses of Syntex's norethindrone (or its acetate), which by then had become the most widely used active ingredient of the Pill.

There is no question that Searle's norethindrone double-bond isomer, norethynodrel, was the first steroid active ingredient of an FDA-approved contraceptive pill and that the company deserves enormous credit for marketing the product in 1960—despite a possible backlash by consumer opponents of contraception. But what about the fact that this substance was synthesized at least a year and a half after Syntex's synthesis of norethindrone and at least a year after my first public report and disclosure of its high oral progestational activity? Given the extraordinary importance of these steroids, why did Colton never disclose any of that chemical work in the peer-reviewed literature?

Colton and other researchers from G. D. Searle had not otherwise been reluctant to publish their steroid work. Why did Gregory Pincus, one of the greatest entrepreneurs and most important figures in the early days of oral contraception, and the person most responsible for persuading G. D. Searle to pursue the commercialization of norethynodrel, make not the slightest reference in his 1965 opus magnum, *The Control of Fertility,* to any chemist (e.g., Frank Colton) or to how the active ingredient of the Pill actually arrived in his laboratory?

One possible reason that Searle's chemical work in this field has never been seen in the bright light of a peer-reviewed journal publication is my pure speculation without hard objective evidence. To my knowledge, this speculation has never been cited before in public, and I have chosen to do so now before all of the active participants are dead. Leon Simon, a respected patent attorney practicing in Washington, DC, since 1945, specialized in the steroid drug field and served from the late 1940s through the middle 1960s as Syntex's outside, independent patent counsel. (In 1965, after the Syntex research and corporate headquarters had moved to the Stanford Industrial Park, Simon followed the company to head Syntex's in-house patent department.) Several years before his death in 1975, he confided to me his own and presumably unprovable supposition of the genesis of Searle's August 1953 patent application concerning norethynodrel.

According to Simon, in January 1952 Dr. Emeric Somlo, then the owner of Syntex S.A., had some negotiations with the Searle family about their possible purchase of Syntex. To facilitate Searle's due diligence examination of Syntex's nonfinancial assets, Somlo instructed Simon to permit the late Dr. A. L. Raymond, Searle's research vice president, to inspect in Simon's Washington office all of the then pending Syntex patent applications. One of these was our Mexican patent application of November 22, 1951, which disclosed the structure, the progestational activity, and the specific experimental details for the synthesis of norethindrone. What, if anything, Raymond did consciously or subliminally with this proprietary information after returning to the headquarters of G. D. Searle and Company in Skokie, Illinois, will never be known.

ORAL CONTRACEPTIVES: PROGNOSIS

Syntex deserves credit as the institutional site for the first chemical synthesis of an oral contraceptive steroid—a statement that is not in any way meant to denigrate Searle's commitment to the contraceptive field and that company's successful drive to be the first on the market with a steroid oral contraceptive. Interestingly, Syntex-developed norethindrone is still one of the most widely used active ingredients of oral contraceptives, whereas Searle's norethynodrel disappeared from the market many years ago, to be superseded by other 19-nor steroids, all of which are close chemical relatives of norethindrone.

Even in the late 1950s people were asking about long-term side effects—what would happen if a woman were to take one of the oral contraceptives for 20 years? To me, this seemed an academic question because I considered it almost inconceivable that any of these synthetic 19-nor steroids would still be in use 20 years hence. Research in this area was in the very early stages, and I was confident that within a few years these compounds would be replaced by others that not only would be more active, but would act by very different mechanisms and at a much lower level than the pituitary–hypothalamic level. Unfortunately, I was completely wrong.

Norethindrone and its relatives are not historical relics; not only were they the only oral contraceptives of the 1960s but they are the only oral contraceptives of the 1990s. That is a remarkable and in many respects a discouraging phenomenon. With the exception of RU–486 and the prostaglandins in chemically induced abortions, no fundamentally new practical discoveries in birth control have occurred during the first 30 years of Pill use; the only new developments have been chemically minor variations on the compound that we synthesized in 1951 and on new sustained-delivery methods of these steroids.

FATHER OF THE PILL

Based on the familial metaphor I proposed at the beginning of this chapter, there is no question that the biologist, the late Gregory Pincus, fully deserves the title "Father of the Pill." It is curious, however, that Pincus never seemed to have had much appreciation for the chemist's role. As I have mentioned, among the hundreds of pages and 1459 references in Pincus's *The Control of Fertility,* no mention is made of a single chemist or chemical publication or how the steroid chemical contained in the Pill actually arrived in his laboratory. The active chemical ingredient did not occur in nature, nor was it bought in a drugstore. Was this omission just a reflection of the low opinion Pincus and other biologists had of the role chemists play in the development of a new drug?

In 1978, at an American Academy of Arts and Sciences-sponsored meeting on the history of birth control in America, I raised this question with Celso-Ramon Garcia, one of Pincus's original collaborators, which led to the following exchange taken from the transcript of the taped proceedings:

> Garcia: Basically, the monograph *The Control of Fertility* that Pincus wrote expressed in detail what his feelings were about who contributed to what.
>
> Djerassi: Why did he not mention any chemists, do you happen to know that?
>
> Garcia: He was a biologist, the same way as you are principally presenting your story as a chemist.

> Djerassi: That's not true; that's why I submitted a paper here with biological references, including yours.
>
> Garcia: Well, okay, but the fact is that principally you are a chemist and your major contribution has been that of a chemist.
>
> Djerassi: But this would be like my describing the history of oral contraceptives without a single reference to Pincus or Rock or yourself!

I shall not dwell any further on the Pill's history, because the chemical portion has been described by me in excruciating detail in my autobiography and the biology, in even greater detail, by Pincus. In some of the subsequent chapters, I would like to expand the definition of the capitalized word "Pill" beyond the narrow meaning of a steroid oral contraceptive, to employ it as a paradigm for any fundamentally new method of birth control—an expanded meaning with which, I suspect, Gregory Pincus would have concurred.

3

Progestins in Therapy—Historical Developments

Hechter: May I take a couple of minutes?

Djerassi: I haven't finished. I'd like to continue because I've only gotten to the first half of my story.

Reed: He can have my time. This is the first really fruitful . . . (inaudible)

Greep: This is history from the horse's mouth, and I think it's very good.

Djerassi: I misunderstood. Did you want me to continue?

Greep: Yes.

The dialogue above is taken from the taped transcript of an unusual session held Friday morning, May 5, 1978, in an old New England mansion on the outskirts of Boston, the headquarters of the American Academy of Arts and Sciences. The "Boston Academy" as it is sometimes known, to distinguish it from its younger counterpart, the National Academy of Sciences in Washington, was holding a closed two-day session on "Historical Perspectives on the Scientific Study of Fertility." Invited were some of the key scientists who had been active in the field of fertility in the United States during the previous 40 years (therefore, it was not surprising that, as far as I could tell, at age 55, I was the youngest of that group) and several much younger historians of science. The purpose of the meeting was to have a free-flowing dialogue

Note: Original text written in 1983.

between these two groups, to collect a record that historians of science might draw upon in the future.

The unedited transcript of that Friday morning session reads awfully: Nouns do not match verbs, tenses get mixed, punctuation is lost, and many words are misspelled or appear to be inaudible. Nevertheless, one gets a real flavor of excited human dialogue and interruptions, of hurt egos, of hitherto undisclosed vignettes; we can hope that some day science historians will do justice to this material.

The scientific co-chairman of the Boston Academy's May 1978 meeting was Roy O. Greep, a distinguished endocrinologist, who has known personally most of the actors in this play. Another key participant was Oscar Hechter, who was senior scientist of the Worcester Foundation for Experimental Biology. He was not directly involved in the development of oral contraceptives, but he was an intimate collaborator of Gregory Pincus. James Reed of Rutgers University, quoted in the epigraph, was studying the birth control movement in America.

I shall emphasize chemistry in my historical presentation, not only because I am a chemist, but also because the chemical history of progestins is relatively unfamiliar to biologists and clinicians. In fact, it is frequently not presented at all.

The chemical history of progesterone and of progestationally active steroids can be divided into four phases. The first phase (circa 1934 to 1940) encompassed the isolation, structure elucidation, and synthesis of progesterone from cholesterol as well as the preparation of a number of simple analogues of the natural hormone to determine what chemical changes were still consistent with retention of biological activity. The second phase (1939 to 1945) is represented by Russell Marker's initial discovery at Pennsylvania State University, and subsequent demonstration on an industrial scale in Mexico, that diosgenin was a convenient and cheap starting material for the preparation of progesterone in tonnage quantities. Although the soybean sterol stigmasterol was later developed by the Upjohn Company as a viable industrial alternative, there is little doubt that Marker's discovery converted progesterone from the status of an expensive rarity to the cheapest of all steroid hormones.

The third phase, in my opinion, began in 1944 with Maximilian Ehrenstein's demonstration at the University of Pennsylvania that removal of the C-19 angular methyl group of progesterone does not lead to diminution of biological activity. Subsequent work by our group at Syntex and by researchers at G. D. Searle Company led to the introduction, in 1957, of both norethindrone and norethynodrel into clinical practice for the treatment of menstrual disorders and, subsequently, as the first oral contraceptives.

The fourth phase, starting in the mid-1950s, encompassed the synthesis of hundreds of analogues of norethindrone and progesterone. This work led to a number of clinically important, orally effective progestational agents.

Until the mid-1940s, virtually all of the progesterone needed for biological and clinical purposes was prepared in one way or another from cholesterol through three of its oxidation products. Because the yield of these various products in the oxidation of cholesterol was poor, it is not surprising that the overall conversion of cholesterol to progesterone was very low and the price of that hormone very high (approximately $80/gram in the early 1940s). All this changed dramatically when Marker revolutionized the chemical production of progesterone. In fact, within a few years, the cost of progesterone dropped so dramatically, as a result of his process, that it became inexpensive enough to be used as the starting material for the synthesis of other steroids, in addition to being a clinically useful steroid hormone. What was the nature of Marker's discovery?

In the late 1930s and early 1940s, Marker performed research on a group of steroids called sapogenins. Marker concentrated on the structure elucidation of diosgenin and, most important, succeeded in developing a four-step, high-yield conversion of that sapogenin to progesterone. Marker was also responsible for discovering that diosgenin was abundant in certain types of yams (*Dioscorea* species) that grow wild in Mexico. Many stories, most of them apocryphal, have been written about Marker's activities, but in 1979 (approaching the age of 80), he visited me at Stanford University and permitted a taped interview. In my opinion, the historical record of progesterone will be better served if I quote Marker rather than paraphrase his recollections despite the raw character of the taped conversation, which was not meant to be published in its original form. In spite of this esthetic deficiency, I believe that it provides a personal flavor appropriate to the type of historical record I wish to describe.

> Djerassi: This is October 3, 1979, and I am finally having the meeting with the great Russell Marker who can tell me what really happened in Mexico. Just tell me that one part over again—you collected about 10 tons of *Dioscorea* in Mexico. . . .
>
> Marker: After I was convinced that Parke-Davis would not go into it, I tried other companies to get support. For

instance, I tried Merck and they said that since Parke-Davis turned me down they could not go into it. . . . Then I decided that I was going to go into it myself and I withdrew from the bank about half of my meager savings and went to Mexico, and I collected 9 or 10 tons of root from the natives that found the original two plants for me. I collected that between Cordova and Orizaba, near Fortin. The man that had collected the original had . . . a little store and a small coffee-drying place right across the street and we collected material and he chopped the material like potato chips and dried it in the sun, and I took it up to Mexico City and had it ground up. I found a man that had some crude extractors there; he extracted it with alcohol and evaporated it down to a syrup. And that I took back to the United States to a friend of mine who had a laboratory, and I made arrangements with him that if he would do the rest of the financing and let me use his laboratory, I would give him one-third of the progesterone that we got. I told him that I expected a little over 2 kilos, I thought. But we ended up with having a little over 3 kilos and he took a kilo of it. At that time he was getting $80 a gram for it.

Djerassi: But how did you carry out the degradation of diosgenin to pseudodiosgenin, which, after all, was really an autoclave reaction?

Marker: Yes, I carried that out in his laboratory . . .

Djerassi: Oh, he had an autoclave?

Marker: Yes, that's right, he had a metal autoclave.

Djerassi: On what scale did you do that first? I am just wondering how much diosgenin did you degrade?

Marker: About 2 kilos at a time.

Djerassi: Had you ever done it on that scale at Penn State before?

Marker: Yes.

Djerassi: How did you meet Somlo? [one of the founders of Syntex and the former principal owner of the small Mexican pharmaceutical firm "Laboratorios Hormona"]

Marker: I went to several people in Mexico with the hope that someone would be interested. I went to the telephone directory while I was staying at the Hotel Geneve . . .

Djerassi: . . . and you looked under hormones?

Marker: No, I looked under "Laboratorios" and I found Laboratorios Hormona and I thought they must be interested in making hormones. So I took a taxi and went out and Lehmann [scientific director and minority stockholder of Laboratorios Hormona] was there. Lehmann looked at me—apparently he thought I was crazy or something, when I first went in and then he excused himself and went out and when he came back he said, "Oh, you are the Marker that has published these papers?" He said it sort of rang a bell that, "I have seen your name some place."

Djerassi: He took you seriously then?

Marker: He took me seriously, and he wanted to know if I would be in town for a few days. He said that Dr. Somlo who owned the company was in New York. So he called Dr. Somlo and told him to come back immediately. And Dr. Somlo came back the next day or so, and I had a talk with Dr. Somlo; they wrote out a small contract that we would start the production and start a new company as soon as I would be available, and things like that. I told them that I had some research that I wanted to finish up before I came to Mexico and that they would finance it.

Djerassi: And did they know that you had made some progesterone in the States before?

Marker: No, I didn't tell them anything about it. So several months later I came back to Mexico, and as I was leaving I told Lehmann that I had made several kilos of progesterone in the States and he was greatly surprised at that and he wanted to know what I had done with that. I told him that I still had it in my possession and so I got back to the States and got a phone call from Somlo; he wanted to know if I still had that progesterone. I told him that I did. He said, "Meet me in New York in a few days," and I told him I would. So he said, "We will set up a company in Mexico for 500,000 pesos," which was a little over $100,000 in those days—pesos were worth roughly 21 cents, and he said, "We make a deal that you are going to have 40% of the stock but you don't have any money to pay for it." He said, "Give me the progesterone and we will start selling the progesterone in this company." So I made a deal with him, and he said we would take out the first $40,000, which you owe for the

40% of the company that we were going to form. It didn't have a name at that time. And the rest of the money that we get for this—see, we were selling for $80 a gram at that time—we'll put into the company's profits and we will split the profits then so that I get 40%, Lehmann 8%, and he 52%.

Djerassi: I just wanted to tape this because what I am finding out is that a lot of the early stories about Syntex are just fiction, and I really would like to hear it for the first time from you. . . . Who thought of the name Syntex? Was that Somlo or you?

Marker: Somlo came in one day when we were about ready to start the company and he said that he had a name for the company—Synthesis he was going to call it. I asked him, since we were down in Mexico, why not have something to indicate that it was Mexico. He said, "All right, Syntex." That's how it started.

Djerassi: Where did you set up a lab there, at Laguna Mayran?

Marker: At Laguna Mayran—it was then the old Hormona building. See, then there was a vacant lot on a corner adjacent to Hormona and during the year they put up some laboratories there for me to work in and a place for the extractors. Well, after a year's time I went to Somlo. In the meantime, I spent all my money and my wife was in Mexico and I had to send her back because of being short of money, and it was cheaper for her to live in State College than in Mexico—even at that time. He would give me enough money to live on: From time to time I would go to him and say, "I am short on money, I need some money to pay my hotel bills and to send to my wife." He'd give me $1,000 or something like that, you see, until I would come to him the next time.

Djerassi: You mean, he didn't even pay you a salary?

Marker: No, no salary, nothing.

Djerassi: Why did you do that?

Marker: The agreement that we had—we were going to split the profits, so at the end of the year, well it was probably in February or March, I went to him and I asked him about the profits because I knew there were substantial profits. I had made at least 30 kilos of progesterone, which I

> turned over to him, and some of it was going to Argentina and was reshipped to Germany during the war. That was another thing I objected to, that the product was being shipped to Germany; at least 2 kilos of it that I know happened that way. Well, I asked him about the profits because the progesterone at that time was still selling for $25 to $30 a gram; and he had reduced the price somewhat, and making some 30 kilos—well, you would have a profit of maybe half a million dollars or something like that. He said, "What profits?" I said, "The profits we made on Syntex." And I asked if I could see the books. And he said, "No, you wouldn't understand them anyway." I told him I would get someone who speaks Spanish to look over the books. He said, "I refuse to let you see them because you couldn't interpret them." Finally he got pretty mad at me and he said, "There is no profit at all." I asked him where the profits were. He said, "I took them as salary and you can't do anything about it." So I decided to leave.

When Marker checked the typed transcript of our taped interview he added the following paragraph, which describes his industrial activities in Mexico after his departure from Syntex and just before he withdrew from chemistry:

"When I left Syntex in May 1945, I formed a company known as Botanicamex in Texcoco and produced progesterone there until about March 1946, when production was moved to Mexico City with Gedeon Richter, who formed a company known as Hormosynth, later changed to Diosynth. While at Texcoco, I produced about 30 kilos of progesterone, approximately the same amount as I produced during my stay with Syntex."

Somlo and Lehmann, looking for another chemist who would reestablish the manufacture of progesterone from diosgenin at Syntex, met Dr. George Rosenkranz in Havana and recruited him to the firm. Rosenkranz had emigrated to Cuba from Switzerland, where he had received his Ph.D. degree under Leopold Ruzicka, one of the giants of early steroid chemistry. Rosenkranz's thesis had dealt with another group of sapogenins, not related to steroids, but he was well acquainted with Marker's publications. Within two years, Rosenkranz reinstituted the manufacture of progesterone from diosgenin, and shortly thereafter also worked out the commercial synthesis of the male sex hormone testosterone from diosgenin—a process that Marker had not implemented at Syntex. By the early 1950s Syntex was producing progesterone at the rate of several tons per year. The reason for this enormous production was not the clinical demand for progesterone, which probably amounted to only a few hundred kilograms for the entire world, but was because it proved to be an ideal intermediate in the synthesis of corticosteroids developed at that time by the Upjohn Company.

4

The Manufacture of Steroidal Contraceptives: Technical Versus Political Aspects

Contrary to predictions made in the 1960s, steroid oral contraceptives played a much larger and ever-increasing role in worldwide fertility control in the 1970s. Indeed, on a global basis oral contraceptives were then equalled or surpassed only by condoms or abortion as the principal components of family-planning programs.

Qualitatively, all of the oral contraceptives have similar side effects because well over 80% are based chemically on either norethindrone or its 18-methyl homologue norgestrel. But in spite of these similarities, there is a small but significant chemical difference (an extra methylene group in norgestrel) which is responsible for the fact that norethindrone, but not norgestrel, can be manufactured by "partial" synthesis—that is, from another steroid intermediate. The problems of steroid manufacture can be separated into two main topics—choice of starting material and selection of chemical steps. Both topics have technical and economic features, but the raw material question is also complicated by significant political considerations, which have so far received very little recognition.

Note: Original text written in 1976.

METHODS OF SYNTHESIS

Norethindrone (as well as Searle's norethynodrel) was first obtained by way of various chemical transformations of estrone. In the 1950s and early 1960s, the starting material for the synthesis of the female sex hormone estrone was largely diosgenin. This synthesis and relatively minor but economically important modifications formed the basis of the initial industrial manufacture of norethindrone and related 19-nor steroids, which was carried out primarily by three pharmaceutical firms: Syntex (Mexico), G. D. Searle (Mexico and the United States), and Schering A.G. (Mexico and Germany).

On an industrial scale, the most cumbersome step was the metal–liquid-ammonia reduction of estrone, and many attempts were made to circumvent it. A general solution was developed independently at Ciba (Switzerland) and at Syntex (Mexico). By the middle 1960s, this second process (still involving diosgenin but not any more estrone) had replaced the lithium–liquid-ammonia reduction in the industrial synthesis of most 19-nor steroid oral contraceptives.

An enormous amount of work was carried out on the total synthesis of steroids—meaning from coal, air, and water—most of it prompted by academic interest. However, the discovery of the high biological activity and the eventual commercial introduction as an oral contraceptive of norgestrel—a steroid unsuited to "partial" synthesis by chemical transformation of a naturally occurring steroid raw material such as diosgenin—immediately raised the question of an industrially feasible total synthesis. Extensive work, notably in the former Soviet Union, the United States, France, and Germany led to practical total syntheses of norgestrel as well as of the more conventional 19-nor steroids (such as norethindrone) that had been accessible by partial synthesis from diosgenin and other plant-derived steroid raw materials.

In addition to such synthetic chemical methodology, microbiological transformations of progesterone and structurally more complex steroids in the 1950s caused an industrial revolution as far as the synthesis of corticosteroids was concerned. One of the reasons that fermentation techniques did not enter more rapidly into the oral contraceptive manufacture was the relatively low price of diosgenin, derived principally from Mexican yams, which for nearly 15 years had constituted the most important starting material. This situation changed markedly, however, and with the exception of norgestrel, during the next few years fermentation techniques had a major impact on the industrial production of most other 19-nor steroid oral contraceptives.

THE ROLE OF MULTINATIONAL PHARMACEUTICAL COMPANIES

The total synthesis of 19-nor steroids such as norgestrel probably represents one of the most complicated and lengthy synthetic processes employed anywhere in chemical industry. Fortunately, the human dose is small (less than 250 mg/year), so that one ton of the final product satisfies the annual requirements of at least five million women. Currently only three Western European pharmaceutical companies are responsible for the bulk of the totally synthesized steroid contraceptives in the world outside of China. Even the complete partial synthesis of a 19-nor steroid is sufficiently complicated, so that only about half a dozen companies in North America, Western Europe, Hungary, and East Germany satisfy the bulk of the world's current demand for oral contraceptives.

In 1975 an American woman buying oral contraceptives with her physician's prescription in a drug store paid an average of $2.75 per monthly cycle, or approximately $1.65 at the pharmaceutical company's level. The price situation was very different, however, if the same oral contraceptive were purchased in bulk in a slightly cheaper but equally efficacious package. In 1975 the U.S. Defense Department purchased five million cycle equivalents at $0.23 per cycle, and the U.S. Agency for International Development let a contract with one pharmaceutical company for 100 million cycle equivalents at $0.1494 per cycle. The profit margin on such bulk sales was so small that it would be very unlikely that any new pharmaceutical company would wish to enter this field. It would be even more unlikely that any newly established government enterprise could produce such quantities at such a low price. These figures are cited to substantiate my contention that for large government-supported birth control programs, the "hardware"[1] component provided by the pharmaceutical companies was relatively cheap, compared with the very much more expensive "software" (e.g., education, distribution, and public health measures) borne by the public sector.

The very few multinational pharmaceutical companies that made the necessary research and production investments in the steroid contra-

[1]I often use the computer terms "software" and "hardware" in the context of fertility control. By hardware I mean the actual methods of fertility control—oral contraceptives, abortion, condoms, coitus interruptus, and so forth. Software elements include the social, cultural, political, religious, legal, and other aspects that are involved in implementation. I have reached the conclusion that any advances in fertility control that take place within the next decade, and probably beyond the end of the century, will concern software rather than hardware.

ceptive field did so only because of the potential return from the affluent markets (i.e., the mid-1970s price of $2.75 per cycle). If it were not for the latter, steroid contraceptives for the large public programs (i.e., $0.15 per cycle) would most likely never have been available.

At times, the differing perspective between the priorities of the private sector (in this case the technically sophisticated pharmaceutical company domiciled in one of the affluent highly developed countries) and the public sector of the less developed country has led to consequences that have all of the marks of a Greek tragedy. For the case of Mexico, I found myself in the role of the Greek chorus, simply commenting on the play "The Rise and Impending Decline of the Mexican Steroid Industry" without being able to affect its inevitable outcome.

SELECTION OF RAW MATERIALS

The key to the nature and even the site of the chemical production of steroid contraceptives resides to a large extent with the choice of the starting material. Total synthesis does not require proximity to any steroid raw material, but in view of its enormous complexity and the need for a large variety of reagents and fine chemicals, production sites are located only in centers of major chemical industry. Both capital and technically trained manpower investments are high.

Once a company makes this major investment, it is unlikely to switch to another raw material, especially one whose availability depends on political or agricultural factors. Partial syntheses can in principle be conducted far away from the site of the raw material production. For instance, hecogenin from sisal is an important raw material for corticosteroid synthesis and even though much of it comes from East Africa or Haiti, it is simply exported to countries such as England and Italy, where all subsequent chemical steps are performed. The same ought to be true of diosgenin—at one time the most versatile of all steroid raw materials—which is extracted from various *Dioscorea* species. The Mexican *Dioscorea* species was particularly rich in diosgenin, but even as early as the 1940s the Mexican government had had the foresight to make the exportation of diosgenin prohibitively expensive. This step resulted in the establishment of an advanced steroid manufacturing industry based on abundant and cheap, locally produced diosgenin in Mexico, initially started by Syntex but subsequently followed by local branches of foreign pharmaceutical firms. In the middle 1950s to early 1960s, well over 50% of all steroids manufactured in the world originated from very cheap Mexican diosgenin; much of it involved either finished goods or at least advanced intermediates. In the early 1970s the Mexican government nationalized the collection of *Dioscorea* plants, and the cost of diosgenin rose so rapidly that by 1975 Mexican-derived finished steroids were often more expensive than those produced by alternative routes.

Like most Third World countries, Mexico suffered from a highly unequal distribution of wealth, which also reflected itself in the availability of medical care and of medicines. It is politically and emotionally understandable that the country wished to do something about this inequity through one of the routes commonly employed: partial or complete state ownership of key industries. Steroid contraceptives cannot be separated from the other steroid medicines (e.g., sex hormones or corticosteroids) because all of them are derived from one starting material—diosgenin. The de facto nationalization in the 1970s of the *Dioscorea* supply and the basic steroid raw material diosgenin was followed by an upward move toward finished goods via a governmentally controlled company "Proquivemex" (Productos Quimicos Vegetales Mexicanos), which intended to manufacture all finished steroids (including contraceptives) used in Mexico, thus attempting to compete eventually with the multinational companies abroad through exportation. Such an "OPEC scheme" will work over the short term only if no alternative raw materials are available and over the long term if no steps are taken to find substitutes. In contrast to the petroleum problem, the situation was very different in the steroid field. Alternative raw materials (e.g., stigmasterol, sitosterol, and cholesterol) rapidly assumed increasing importance, and most importantly for steroid contraceptives, total synthesis, which is independent of any natural steroid raw material, became competitive in view of the extraordinary rise in the cost of diosgenin.

The short-term effect of the nationalization process seemed politically attractive: providing a higher income to Mexican peasants collecting the *Dioscorea* plant raw material and a lower cost of the final steroid drug to the Mexican consumer. However, because of smaller volume production the state-supported enterprise was bound to be much less economic in trying to fill Mexico's internal requirements. The reason for this is that only about 10% of the steroids produced in Mexico by multinational pharmaceutical companies were used in Mexico, the rest being exported. As the cost of diosgenin-derived steroids rose, the attractiveness of Mexican diosgenin dropped precipitously and soon reached a level when the multinational pharmaceutical companies ceased most advanced steroid manufacture in Mexico, shifting production elsewhere via alternative methods and raw materials. The eventual monetary cost (loss of exports and local employment; increased cost of locally consumed steroids) paid by Mexico for short-term political expediency was high and irreversible. The question may well be asked why the Mexican government and the large steroid manufacturers could not have found a mutually acceptable *modus vivendi.* But that is a political question outside the scope of a chemical Greek chorus to this tragedy.

5 Birth Control after 1984

As we enter the decade of the seventies, it behooves us to consider what some of the future contraceptive methods might be and especially what it might take, in terms of time and money, to convert them into reality. There are many publications on this subject, but to date none seems to have concerned itself with the logistic problems associated with the development of a new contraceptive agent. In that connection, it is instructive to note that only total nuclear or chemical–biological warfare receives higher ratings of concern than the problems arising from the world's burgeoning population, and that, of the four top priority problems (total annihilation, destruction of biological and ecological balance, administrative management of communities and cities, and large-scale distress caused by transportation problems and crime), only fertility control requires experimentation in humans for its ultimate solution.

The surprisingly rapid acceptance of intrauterine devices (IUDs) and of steroid oral contraceptives in many developing and developed countries is principally due to the fact that their use separates, for the first time, contraception from copulation, and it is clear that effective birth control methods of the future must exhibit this same property. I have selected three topics relevant to the outlining of logistic problems, the determination of time and cost figures, and recommendations for

Note: Original text written in 1970.

implementation. (I exclude mechanical devices such as improved IUDs for the following reasons: Until now, clinical research with IUDs has fallen outside the scope of government regulatory agencies such as the Food and Drug Administration [FDA]. However, it is highly likely that public as well as scientific pressure on government regulatory bodies will require that such devices also be brought within the scope of their control and that clinical use of these devices be preceded by the same type of stringent testing that is demanded for contraceptive drugs. I emphasize these arguments only to point out that the cost and time estimates made later in this chapter in connection with new chemical contraceptive agents probably will also apply to new devices of the IUD type.)

1. A female contraceptive, consisting of a once-a-month pill with abortifacient or menses-inducing properties. Such a method is scientifically feasible, it should lend itself to use in both developed and developing countries, and it addresses itself to the critically important subject of abortion.

2. A male contraceptive pill.

3. A draconian agent, such as an additive to drinking water. I include this approach not to justify the Orwellian overtones of this article's title, but rather to place into realistic perspective the problems of developing such an agent, which is mentioned with increasing frequency as the final solution if voluntary methods should fail.

Many advances in fertility control considered by the World Health Organization are based in one way or another on chemical approaches. This type of research on fertility control is exceedingly complicated in both preclinical and clinical phases; the required manpower and financial resources are available only in the technologically most advanced countries; and the results of such research are subject to scrutiny and approval by regulatory bodies in such countries. As the FDA has such a crucial de facto power in many foreign countries, it is realistic to base time and cost estimates on the American milieu, where the bulk of human fertility control research is being conducted at present.

FDA REQUIREMENTS AND ANIMAL TOXICITY STUDIES

Irrespective of the sponsor (whether industrial, governmental, or academic), no new drug can lawfully be administered to humans in the United States without an investigative new drug (IND) exemption issued by the Food and Drug Administration (FDA). Appropriately, animal toxicity data must first be presented and, for drugs outside the field of contraceptives, the FDA's requirements in this regard are reasonable; in particular, the choice of the experimental animal is left to the discretion of the investigator.

However, different FDA requirements exist[1] for contraceptives (whether steroids or nonsteroids), and these must be taken into consideration in any time and cost estimate for new fertility control agents. In contrast to the requirements for noncontraceptive drugs, where the animal species is not specified, contraceptives must be tested in rats, dogs, and monkeys.

Nobody can dispute the wisdom of the requirement for data on toxicity in animals before a drug is administered to humans, even in short-term clinical experiments involving only a few individuals. Nevertheless, stipulation of the animal species to be used is extremely unwise. After all, the sole reason for selecting any animal is to provide a model for extrapolation to the human. The unfortunate choice by the FDA of the dog as one of the required species for testing oral contraceptives has already resulted in the suspension of clinical experimentation with three contraceptive agents, notably the chlormadinone acetate "minipill." Indeed, even the simple requirement for data on toxicity in the monkey may be close to meaningless in the area of reproductive physiology unless careful attention is given to the choice of the monkey species.

To gain as much knowledge as possible from animal studies, a species should be selected that most resembles the human in its metabolic handling of the drug in question. (Table 1 illustrates the enormous differences among different animal species in the metabolic handling of one drug, in this instance the anti-inflammatory agent naproxen. In this case, the differences between rhesus and capuchin monkeys are almost greater than the differences for any other two animal species.) Much more work needs to be done in identifying useful animal models that have some predictive bearing on the human biological response to a given agent. Such work will require major efforts on the part of investigators, major financial inputs (notably into primate facilities) and, most importantly, some relaxing of the FDA requirement for rat, dog, and monkey.

[1]These were introduced in 1969 and only changed in 1988 when virtually all American pharmaceutical companies had long withdrawn from significant R&D in contraception.

Table 1. Data on Excretion Patterns and Plasma Half-Life for an Experimental Drug

Species	Excretion		Plasma Half-Life[a] (hours)
	Urine (%)	Feces (%)	
Human	94	1–2	14
Rat	90	2	4–6
Guinea Pig	90	5	9
Dog	29	50	23–35
Rhesus monkey[b]	90	2	2–3
Capuchin monkey	45	54	20
Stump-tail monkey[b]	40	60	1
Mini-pig	86	1–2	4–7

[a]The plasma half-life is the length of time at which only half of the initial amount of drug will be detected in the plasma.

[b]These two species belong to the same genus *(Macaca)*.

SOURCE: E. Forchielli, R. A. Runkel, M. D. Chaplin (Syntex Research, Palo Alto, Calif.), private communication.

ROLE OF THE PHARMACEUTICAL INDUSTRY

Except for certain biologicals (special vaccines), essentially all modern prescription drugs were developed by pharmaceutical companies. I know of no case in which *all of the work* (chemistry, biology, toxicology, formulation, analytical studies, and clinical studies) leading to governmental approval of a drug (e.g., by the FDA) was performed by a government laboratory, a medical school, or a nonprofit research institute. This does not mean that many of the basic discoveries leading to the development of a drug ultimately used by the public are not discovered in such nonindustrial laboratories, or that certain important steps (e.g., much of the clinical work) are not performed outside of industry.[2] Nevertheless, it is a simple fact that, in modern industrial nations, pharmaceutical firms play an indispensable role in the development and distribution of any drug.

The public and legislators are frequently unaware of this key function of the creative elements of the pharmaceutical industry. This func-

[2]Norplant (a sustained-release formulation of an oral contraceptive steroid) was developed by the Population Council, yet its introduction to the public around 1990 required the intermediary of a pharmaceutical firm (Wyeth).

tion is not solely related to the marketing function of these firms (indeed, some pharmaceutical companies do no research but simply acquire their products from other companies). Rather it speaks for their *unique ability* to organize, stimulate, and finance multidisciplinary R&D covering the entire gamut of the scientific disciplines required in converting a laboratory discovery into a practical drug. In addition, the organizational efforts involved in preparing a complete new drug application (NDA) in the United States are completely outside the capabilities of nonprofit institutions and are not undertaken by government agencies, although the latter could presumably mobilize the requisite manpower and funds for such purposes.

Some of the special requirements that have been imposed in the case of drugs used for fertility control are understandable and justified; similar requirements would undoubtedly be imposed in the case of any other drug (e.g., preventive medication in atherosclerosis) administered for long periods (usually years) to normal populations. These requirements are a response to our gradually increasing knowledge of human reproductive physiology in general, our accumulated experience with oral contraceptives in particular, and especially the surprisingly rapid acceptance by so many women of these new birth control agents.

Unfortunately, the costs of developing such agents have escalated to such an extent that it is unlikely that the traditional course of drug development will lead rapidly, or even eventually, to the creation of fundamentally new contraceptive agents. If the present climate and requirements had prevailed in 1955, oral contraceptive steroids would still have been a laboratory curiosity in 1970. It is obvious that toxicity and testing requirements will become more stringent and time-consuming, not less so; other criteria (e.g., tests of potential mutagenesis and more sophisticated metabolic studies) will be added as logical consequences of accumulated new knowledge. Costs are bound to escalate enormously.

BIRTH CONTROL DEVELOPMENTS

Contraceptive methods have been designed for the female not just because she is more receptive to new approaches, presumably since unwanted pregnancies affect her much more directly than they affect the male, but because our knowledge of the female reproductive cycle provides more hints about rational approaches to contraception than our knowledge of the male process does. Furthermore, it is possible to interfere with the female cycle at many more stages. As an important example of future contraceptive methodology in the female, I have chosen a once-a-month pill with menses-inducing or abortifacient properties, or both, because such an agent has at least four advantages over agents now being used.

1. Administration of one pill a month is clearly more convenient than daily administration of pills. This is true both for major fertility control programs in developing countries and for highly motivated individuals in advanced countries.

2. Periodic short-term administration of a drug may be expected to give rise to fewer long-term side effects, primarily because the agent is intended to act more specifically on a well-defined biological process.

3. Because the agent will interfere in one way or another with progestational function irrespective of whether fertilization has or has not occurred, it does not matter whether the woman is pregnant.

4. Ideally, the agent might be active any time during the first eight weeks after fertilization, so that it could also act as an early abortifacient.[3] In case of drug failure, another agent should be available for subsequent chemical abortion, or else surgical termination of the pregnancy should be available as a backup measure.

The importance of abortion as a means of population control has been emphasized many times. In areas of the world (Japan and eastern Europe) where population growth was reduced dramatically within a short period, this was done principally through surgical abortions. Clearly, the availability of a chemical (that is, nonsurgical) abortifacient would be highly desirable.

A critical path map (CPM) for the development of such an agent is shown in Figure 1. Two major comments are required for a full evaluation of this chart. The first refers to the teratology studies, which are extremely important in any agent affecting embryonic development. The unsupported assumption is made that the FDA would permit phase I clinical studies without previous teratology studies in animals. Irrespective of the correctness of such an assumption, such studies and the subsequent phase II and phase III clinical research can be performed only in a location where, in case the method fails, surgical abortion can be employed. Indeed, the work leading to eventual determination of the clinically effective dose will require progressive lowering of the dose until a level is reached in which failure is observed. From an investigative standpoint, it would be desirable if human pregnancies resulting

[3]At the time (1969) these words were written, research on RU-486 had not yet started.

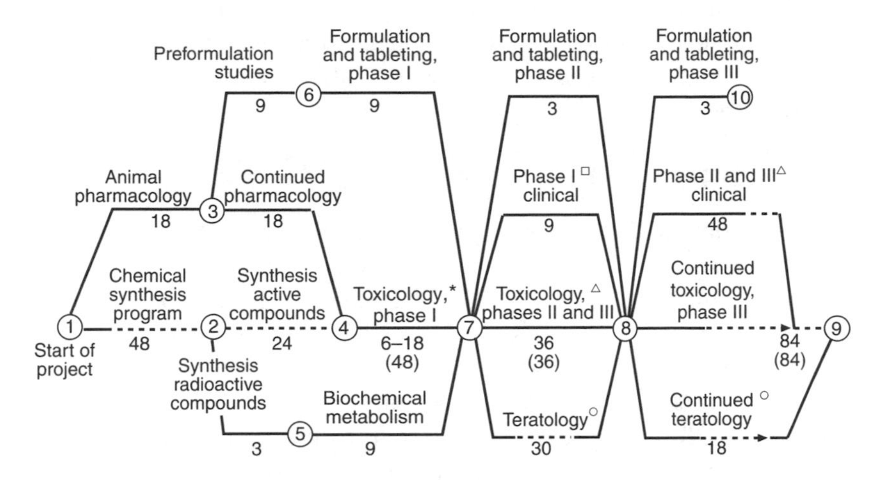

Figure 1. Basic critical path map for the menses-inducing or abortifacient agent. The circled numbers are step numbers; the numbers below the line are time periods, in months. Thus, for example, ①$\underline{\quad 18 \quad}$③ means that the period from the beginning of step 1 to the beginning of step 3 is 18 months. Numbers in parentheses indicate time periods, in months, when the usual FDA toxicological-study requirements for contraceptives are a possible alternative.

from such drug failures were permitted to proceed beyond the 14th week before surgical abortion was undertaken, so that the fetus could be examined for evidence of malformation. This requirement would be difficult insofar as availability and cooperation of patients is concerned. In the absence of such cooperation one would have to depend on monkey data, which are obviously less informative.

The second comment on Figure 1 pertains to the time estimates. These are ideal figures, and the aggregate of about 126–210 months may not be realizable, because it involves almost perfect coordination and even telescoping of various steps in the CPM scheme. For instance, the preliminary toxicological studies (Figure 1, steps 4–7) on 25 compounds will involve rejection of several compounds because of serious toxicity, as well as rejection based on phase I clinical data (steps 7–8). The estimate of 6–18 months for the time required for the initial toxicological studies leading to the selection of the final compound is, therefore, very optimistic. In any event, this time analysis (using 1970 as the starting date) offers the first justification for the title of this chapter, since the middle of the 1980s is already an optimistic target date even when one ignores the time required for the new agent to receive the final stamp of government approval (under current regulations) and be disseminated to the public.

MALE CONTRACEPTIVE AGENT

The condom and withdrawal prior to ejaculation are the only reversible contraceptive measures that are currently available to the male. As has been pointed out by the World Health Organization scientific group[4], "an agent that could safely and effectively inhibit fertility in the male, without risk of interfering with spermatogenesis and libido, would find practical application in fertility regulation." The report then proceeds with a long list and associated bibliography of chemical agents that have been shown to have some effect on the fertility of male animals, notably rats, and concludes, "none of the chemical agents is suitable for use in man, owing to known or potential toxicity. Similarly, immunological processes present hazards when used in man, and they suffer from a lack of specificity. *Consequently, no systemic method of fertility control in man is available at present*" (italics mine).

The CPM chart (Figure 2), therefore, contains a longer estimate than those of Figure 1 for the time needed for discovery of suitable leads that may give rise to compounds warranting clinical investigation. It would be highly desirable if several programs of the type outlined in Figure 2 under steps 1→2→4→8 and 1→3→7 were instituted in several laboratories at the same time in order to increase the chances that a useful agent might emanate from such research. Nothing will stimulate future research on a practical male contraceptive agent more than the discovery of viable and significant chemical leads, but, even in that event, in 1970, 1984 appeared to be an exceedingly optimistic target date for development of a male contraceptive pill ready for use by the public.

Three other difficulties associated with the development of a chemical contraceptive drug in the male must be recognized. First, our basic knowledge of the reproductive biology of the male is even less advanced than our knowledge of that of the female, and a great deal of fundamental work needs to be done, much of it probably in subhuman primates.

Second, the actual clinical work has so far not drawn the attention of planners in the birth control field. The human spermatogenic cycle,

[4]"Developments in Fertility Control," *World Health Organ. Tech. Rep. Ser. No. 424* (1969) (report of a WHO scientific group).

from spermatogonium to ejaculate, lasts approximately 12 weeks. It is likely that testing, including preliminary treatment control and post-treatment recovery observations, might last up to six months, depending on the point where the drug in question attacks this sequence. Pilot testing could presumably be carried out in groups of five to ten males, at each of three widely spaced dose levels for each agent.

Observations should combine evaluation of the effect on spermatogenesis or sperm motility, or both, with observations of organ toxicity and other side effects. At present there appear to be available, in the entire United States, facilities for evaluating only two drugs at a time. Women can easily be assembled for clinical studies through their association with Planned Parenthood clinics and individual obstetricians or gynecologists; there exists no simple mechanism for assembling similar groups of males for clinical experimentation. Ignoring, for the sake of simplicity, important ethical reservations, the prisons and the armed forces are the only convenient sources; thus, results would have to be based largely on examination of masturbation sperm samples rather than on an evaluation of fertility control in an average population.

This leads to the third difficulty—namely, the male's generally lesser interest in, and greater reservations about, procedures that are aimed at decreasing his fertility. If the agent were to be administered orally, men

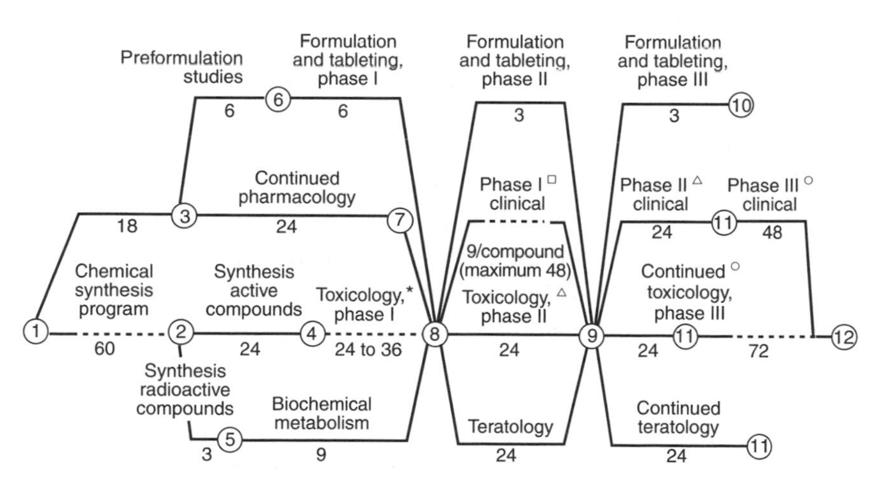

Figure 2. Critical path map for development of a male antifertility agent (*see* legend to Figure 1).

would probably be even less reliable about taking a tablet regularly than women have proved to be, and efficacy could probably be determined on a large scale only through long-term studies of married couples.

The single greatest objection to the oral contraceptives now being used is to the essentially continuous administration of a potent agent to fertile women for many years. Clearly even greater objection would be raised in the case of a male contraceptive pill if it had to be taken day after day by fertile males for many years, possibly 40 or more. However, if both a female and a male contraceptive pill were available, then the two partners could alternate (say every six to twelve months) in their use of a pill and thus avoid continued exposure to one agent for long periods. Such a regimen is likely to work only in educationally advanced and highly motivated groups, and it is probable that the female partner would bear the principal burden of enforcing it.

ORWELLIAN APPROACHES

Some laymen, legislators, and scientists concerned with the economic and environmental effects of rapid population increase have started to imply that drastic government-imposed birth control procedures may have to be introduced if voluntary use of conventional methods fails to stem the tide. I would like to use the adjective Orwellian for such externally imposed extensions of voluntary fertility control. Clearly the most all-encompassing and frightening concept is the addition of a temporary sterilant to water or staple foods. I would like to consider briefly some of the practical problems associated with the development of such an Orwellian agent, which reduce the concept to an absurdity, even if one were to ignore the enormous ethical problems and political hurdles.

1. The substance would have to be active in either the male or the female, but only in their reproductive years, and active over an enormous dose range, because food and water intake of, say, a 20-pound child and a 200-pound adult are very different. It would have to be tasteless. It must be specific for humans.

2. If added to food, the substance would have to be incorporated by the supplier rather than by the consumer in order to ensure universal administration. Even then, a dissenter could simply eliminate a given food from his or her diet and thus escape the contraceptive effects, unless it were a food that is universally required (e.g., salt). In any event, the contraceptive additive would have to be stable during pro-

cessing (baking, heating, or sterilization) and during exposure to oxidants or light in the course of packaging and shipping.

3. Because everyone must drink water, this would seem to be the better vehicle for the contraceptive agent, but even here there would be a difficulty; incorporation would be feasible only when water was supplied through a central system, not obtained from wells. This limitation alone would probably make the method unworkable for at least half the world's population. However, regardless of the method of incorporation into the water, the contraceptive agent would have to display chemical stability on coming in contact with pipes and other metal objects; stability on exposure to light and oxidants in a holding tank or reservoir; stability on exposure to extreme temperatures during cooking or refrigeration (that is, lack of precipitation from solution); no chemical interaction with minerals in the water, and with commonly consumed foodstuffs during cooking; and no properties that would cause problems of over- or under concentration during food processing, as in the preparation of frozen juice or soup concentrates.

4. The question of side effects is insoluble. No drug is devoid of side effects and, in this particular instance, the side effects of the agent would have to be minimal not only in the sex and age group in which it was supposed to be active but also in all other age groups and in the opposite sex. In contrast to any drug now used by humans, which generally is simply a contaminant of the person's microecology, the Orwellian contraceptive added to food or water would be a general environmental pollutant. It would have to be considered a pesticide, albeit one that is directed primarily at humans. It is exceedingly unlikely that such a compound active in humans would be ineffective in at least some other animal species. In fact, because initial biological screening for such an agent would be carried out not in humans but in animals, *an agent truly specific for humans would completely escape detection.*

5. If such an Orwellian contraceptive were completely effective, then its effects would have to be reversible through the administration, presumably by license, of a second agent. The likelihood of discovering such an agent is slight, yet its availability is an absolute prerequisite for employment of the sterility agent. The other alternative would be to develop a contraceptive that significantly reduced but did not abolish fertility, the level of escape then setting the birth rate. Such a property might make such an agent acceptable from a demographic, but hardly from a personal, standpoint.

In the light of these special problems, which would have to be superimposed on the already formidable difficulties associated with the development of any systemic, chemical agent of fertility control, it is perfectly clear that the development of such a universal birth control agent is outside the realm of possibility for many decades to come.

In spite of their inherent scientific plausibility, practical immunological approaches, though more easily implemented in an Orwellian society than the addition of a sterilant to food or water, are still far away. We are thus brought back to reality with only two methods that could conceivably be introduced on a massive scale by government edict during the next two decades. In the male, this would be vasectomy (essentially an irreversible procedure), and in the female, administration of a sustained-action contraceptive of the estrogen–progestin type.

GENERAL RECOMMENDATIONS

The inevitable conclusion is that the pharmaceutical industry ought to remain involved in the massive effort required to bring a fundamentally new female or male contraceptive agent to fruition. Furthermore, most of this work has to be, and will be, done under rules and regulations established by the FDA and similar government regulatory agencies of the technologically most advanced countries. If this premise is granted, then the following four recommendations should be taken into consideration, with the first two constituting needed stimulation of contraceptive research irrespective of what organization (industrial, governmental, or academic) sponsored the drug trials.

1. *Conditional approval.* The FDA (and government regulatory agencies in other countries such as the Food and Drug Directorate in Canada) has two principal functions that are potentially conflicting.

The first function, which clearly should not be abolished, is that of protector of the consuming public insofar as drugs on the open market are concerned. The FDA must protect the consumer from harm and fraud, it must maintain and enforce appropriate analytical standards, and it must generally assume the function of policeman or watchdog. This historical function of the FDA is at least partly incompatible with its role in passing on all clinical protocols by having a de facto veto on all clinical work with experimental drugs. It is at this premarketing stage of a drug's development that the maximum flexibility commensurate with scientific caution and medical responsibility must be maintained; the agency responsible for such protocols must consider its main function to be stimulation of research and drug development rather than just a policing function.

Thus, the role of the FDA seems to have moved from that of protector to that of guarantor; Congress, the press, and consumer protective groups are responsible. Yet it must be recognized that this role of guarantor is an impossible one. No drug can be *totally* effective and *completely* safe, and no agency of government can guarantee that it will be.

The consumer also suffers from the delusion that drug safety and drug efficacy are all-or-none propositions. The fact that people experience side effects from "safe" drugs should be no more surprising than the fact that occasionally some people die when "safe" airplanes crash. This evaluation leads to the following recommendation for a change in procedure that may be beneficial in facilitating and stimulating research not only on contraceptive drugs but also on other drugs in preventive medicine involving long-term administration to "normal" populations.

For such drugs, the IND/NDA process as it exists is totally inadequate and should be modified. The existing phase III clinical program should be reduced to meticulously planned moderate-sized clinical studies of limited length (two years would be adequate in most instances), which would disclose whether a new agent had any conspicuous toxicity. Efficacy could clearly be established under such conditions. The question of whether the drug had any low-incidence toxicity would remain. The oral contraceptives have taught the medical profession the important fact (well known to statisticians) that large samples are needed to demonstrate small effects reliably and that it is extremely difficult and costly to accumulate such samples in a premarketing phase.

At this stage the FDA could introduce the concept of *conditional approval*, somewhat analogous to the FAA's "Certificate of Provisional Airworthiness." Conditional approval means that during use-testing the agent could be marketed, but some of the profits from sales would be used for structured follow-up studies of sizable populations consisting of the patients put on medication. The FDA could assign a permanent monitor to co-administer such programs.

One would avoid the need to collect, during clinical trials, tremendous quantities of information on people who are well and reacting favorably to the drug. Instead, attention would be focused during the "provisional-approval-for-marketing" phase on the few individuals who did poorly, and it would be possible to determine more quickly whether their reactions were drug-related. If the drug survived a well-designed follow-up study, then it could be given full approval by the FDA, and continuing large studies financed by the sponsor would not be required. Implementation of such a recommendation could markedly speed up the time required to develop a practical contraceptive agent.

2. *Appeal procedure.* All clinical research performed in this country is subject to disapproval by FDA personnel. Disapproval for the initiation or continuation of clinical trials is essentially unappealable, and yet such action is frequently a result of hypercaution rather than of exceptional scientific insight. A procedure for appealing such scientifically debatable decisions is urgently required in the field of birth control, because lack of the right to appeal is already having serious repercussions in the form of discontinuance of major research projects.

3. *Patent protection.* Consideration should be given to a possible revision of the patent lifetime of drugs in the area of birth control and in other fields (e.g., vaccines) where very-long-term, premarketing investigation is required. At present the life-span of a U.S. patent is 17 years. Clearly, if a pharmaceutical firm invests tens or hundreds of millions of dollars in research over a period that consumes most of the lifetime of the patent (a circumstance that may easily happen when a 10- to 15-year period of premarketing research and development is required), then a crucial incentive is removed. One possibility is to offer use-patent protection for such products for, say, 10 years, starting with the date of the *approved* NDA.

4. *Government–industry interaction.* The costs of developing a new contraceptive agent have risen dramatically. The reason for these tremendous costs and for the long experimental periods is the readily understandable one that a drug administered to large portions of the normal population must present minimal risk. The chances of developing such drugs are correspondingly smaller than those of developing other drugs, and it is only reasonable that the public (that is, the taxpayer, by way of the government) should bear part of this development cost.

The very special features responsible for the extraordinary costs of birth control drugs are the very long trials required to determine toxicity (completely unlike those for other drugs and eventually concentrating largely on subhuman primates) and the very large and long clinical trials in humans, accompanied by an ever-increasing number of clinical laboratory examinations. This aspect of the research, rather than the chemical, biological, short-term toxicological, or even certain clinical studies, should be funded by the public in the following manner.

A pharmaceutical company should be given the option of applying to a government agency for full financial support of the long-term toxicity studies (which could actually be performed elsewhere under contract) and of all phase III clinical work. If the research should lead to development of a commercial product, then the company would be obligated to repay the government agency on an annual royalty basis. If all of the money were repaid and the drug were still being sold commercially, it might be reasonable to expect a continued royalty payment on a reduced basis for the life of the commercial product. In other words, during the first years of such a system, funds would only be outflowing from the government agency, whereas after a certain period an equilibrium would be reached. Under extremely favorable circumstances the flow might even turn in favor of the government agency.

Such a proposal may appear unprecedented in the drug field, but it had a striking precedent in the U.S. government's decision to underwrite the development of a supersonic transport (SST) in this country. The socially redeeming features of the SST cannot compare with those of a drug in the birth control field, nor are the respective effects of these developments on the environment in which we live comparable. Expenditure in the birth control field of the monetary equivalent of one or two SSTs per year could have an astounding effect and, at the same time, could serve as an indication of how national priorities should really be handled.

My fundamental purpose in making this proposal is not to argue the advantages of the free enterprise drug industry or to protect its profits. It is to ensure the continued possibility of the development of effective contraceptives that are vital for human well-being. To ensure this we must decide either to create an effective partnership between government and industry, on the model of other major technological efforts such as the space program, or to undertake the difficult and even more costly steps that would be involved in socialization of the drug industry in areas requiring long development periods.

I end by predicting that fundamentally new birth control procedures in the female (e.g., a once-a-month menses-inducer or abortifacient agent) and a male contraceptive pill probably will not be developed until the 1980s at the earliest, and then only if major steps of the type outlined here are instituted in the early 1970s. Development during the next decade of practical new methods of birth control without important incentives for continued active participation by the pharmaceutical industry is highly unlikely. If none is developed, birth control in 1984 will not differ significantly from that of 1970.

Future Prospects in Birth Control

6 Reversible Fertility Control

In 1971, a discussion of chemical approaches to fertility control could be presented in several ways. One would be to emphasize the indispensable role that chemists have played and are playing in the development of contraceptive agents. Some of the chemistry of biologically intriguing compounds is trivial (e.g., 3-chloro-1,2-dihydroxypropane), but much of it is elegant, fascinating, and stimulating. One need only cite the multitude of total syntheses of steroids; the sophisticated chemical steps developed for the transformation of naturally occurring steroids to 19-nor steroids; the isolation, structure elucidation, and synthesis of the prostaglandins; or the chemistry of the releasing hormones of the hypothalamus to complete the case for the chemist's past and present creative function in fertility control.

The second approach could consist of painting in broad outline a picture of the promising and even sensational advances in fertility control that are just lurking around the corner—immunological approaches, male contraceptive pills, once-a-month pills or once-a-year implants in the female, additives to food and water, and so on. I am afraid that the scientific and lay press are replete with such optimistic predictions and that, as a result, scientists and laymen alike are expecting sensational advances that, in my opinion, are totally unrealistic from an operational and time standpoint. The basic premise behind these optimistic predic-

Note: Original text written in 1971.

tions is perhaps defined most concisely by the following typical quotation, in 1969, by one of the clinicians involved in the very early (1956) studies with the Pill: "The agents available today will undoubtedly give way to better ones, for the ingenuity of man is infinite, and untold technologic resources remain to be tapped by his wisdom." Once this dubious premise is accepted, then it does not become too difficult to accept predictions, albeit unsupported ones, of the following types, which appeared in the technical or lay press between 1968 and 1971:

"A shot to immunize women against pregnancy, a pill to render men temporarily sterile, and a monthly pill to ensure a regular monthly flow may all be in common use within the next decade."

"The contraceptive pill for men may only be four or five years away, depending upon the level of funding and political attractiveness."

"The prostaglandins, more than anything else now in sight, herald an entirely new generation of contraceptives, a generation that is destined to make the Pill seem as dated as the diaphragm, as clumsy and unappealing as the condom. Every woman can now look forward to simpler, safer, surer contraception in the years ahead."

"Given the current interest in the study of reproductive physiology and methods of contraception, I think it is very likely that in a relatively short time—five to fifteen years—scientists will discover ways of controlling the fertility of an entire population."

A third presentation could be grim, and I have chosen this alternative for this chapter. Population pressures are rising precipitously (e.g., Pakistan's birth rate is such that the 500,000 deaths from the November 1970 cyclone were already "made up" in January 1971), but in spite of the chorus of optimistic predictions, no fundamentally new practical birth control agent is in sight that could be used in 1972, 1973 . . . 1975 to stem the tide more effectively than presently employed procedures. In the highly developed countries, notably the United States, there currently exists a climate of scientific retrenchment that, for preventive medicine (i.e., fertility control), is now so compounded and interwoven with bureaucratic hypercaution and virulent consumerism that it is unlikely that major new advances in birth control can be brought to practical fruition in the United States in this decade. The present climate penalizes the imaginative approach and encourages minor and "safe" modifications of existing methods. In retrospect, it is clear that the most significant postwar development in birth control—the steroidal oral contraceptives—could not have been realized in the present climate.

Most, if not all, of the theoretically feasible, new fertility control agents are chemical in nature, and one might expect, therefore, that chemists would play a decisive role in their development. Although this was true with many conceptual and operational aspects of the creation of oral contraceptives, the chemist's role is likely to be much less dominant in future contraceptive developments. The pitfalls and barriers in drug development, notably in drugs involved in preventive medicine, are now largely nonchemical in nature. The most serious deficiency—the lack of knowledge and the relative unavailability of suitable animal models—is particularly restrictive in reproductive biology research because reproduction is the most species-specific biological property, and extrapolation of results from lower mammals to humans is fraught with danger. Yet the panic responses of regulatory agencies and of the press in the highly developed countries are most frequently predicated on experiments with doubtful, or possibly even meaningless, animal models. The withdrawal of cyclamates or of certain oral contraceptives from the market at the demand of regulatory agencies was based on experiments with rodents and beagles, respectively, which were carried out many years *after* these agents had been used successfully in humans. It is, of course, conceivable that these decisions were correct because it is impossible to produce ironclad evidence that such animal experiments are totally meaningless. Once this is granted with existing agents that have been used for long periods of time in humans without apparent deleterious effects, then one must obviously employ even more caution with agents that are only in the research stage.

Precisely this dilemma is currently being faced in fertility control research—there are exciting leads in the laboratory, but it is becoming a sisyphean effort to convert them into practical realities, namely, agents that can be administered to millions of people. This explains the apparent contradiction between the pronouncements of the prophets of a glorious future in birth control and the prophets of gloom. The exciting advances in the laboratory announced by first-rate scientists are mostly glorious platitudes about impending advances in human fertility control because most of these honest enthusiasts are totally unaware of the logistic problems that need to be faced before a drug or device can, and should, be employed as a practical fertility control agent.

From a global viewpoint, the fantastically accelerating growth rate of the world's population (in the year 1800 it took 100 years to produce an additional billion people, but at the present time it will take only about one dozen years to accomplish this feat) is probably the most serious problem facing us in the immediate future. However, if we ignore ecological and environmental problems, then the population problem in the technologically advanced countries is not a matter of sheer survival, whereas it clearly is that in the lesser developed countries. Put into other terms, North America and Asia have approximately the same land surfaces, but although North America contains only about 10% of the world's population, this minority has nearly half of the world's income; Asia boasts of over half the world's people, but they earn only about 10% of the world's income.

Much earlier, I called attention to the implications of this situation in the field of fertility control research. Such research is enormously complicated and costly—both in terms of manpower and money—and most of it is performed in the most highly developed countries of Western Europe and North America. These areas of the world have the greatest affluence and the smallest population growth rate. It is not surprising, therefore, that the citizens, the press, and the regulatory bodies of these countries now impose extraordinary, and perhaps even unrealizable, safety requirements on new birth control agents. Existing birth control agents, though imperfect, do cope with their existing birth rates, and other factors, therefore, get a higher priority rating.

The priorities in lesser developed countries are completely reversed, but their governments face the double dilemma of not having the financial and technical resources for extensive fertility control and development work and yet of having to accept the "safety" standards of the highly developed countries for simple political expediency. Even more pernicious are the implications of clinical work performed in lesser developed countries with agents that have been discovered in highly developed countries but not permitted to be tested there because of regulatory restrictions. Looked at in this light, the following words of a Chilean scientist seem surprisingly mild: "In the past, several countries have been accused of being pawns to 'Yankee imperialism,' and Chile has been said to be an experimental field for Americans to find out how much can be obtained in matters of birth control in an underdeveloped country." Until now, relatively few institutions or investigators in the advanced countries have set themselves the objective of creating fundamentally new fertility control agents that are specifically designed for important population sectors of the lesser developed and more rapidly breeding countries. Considerations that ought to enter into the design of such tailor-made contraceptive agents are not just scientific in nature; frequently, political, economic, sociological, religious, and other cultural factors are likely to play an even more significant role, but, to date, they have either been ignored or else have been considered at much too late a stage.

The Chilean demographer Requena has drawn an oversimplified, but nevertheless very useful model (Figure 1) to relate socioeconomic–cultural levels with birth control practice. This model was done in the context of Chile, but it does have much broader validity, and I have selected it as the vehicle for assigning relative priorities to various birth control approaches. The main thesis of Requena's argument is that contraception constitutes an aspect of preventive medicine—an argument with which I concur heartily—and that the conscious use of preventive medication requires a fairly high cultural level. The lower the cultural level, the more likely it is that the person will act after the consequences (illness, pregnancy, impending death, etc.) have become obvious. It follows that at the lowest socioeconomic and cultural level, one encounters (a) the highest number of births, (b) virtually no use of contraceptives, and (c) a minimal incidence of induced abortion. As one progresses on the cultural ladder, the number of births declines and the number of abortions increases dramatically because, at this stage, the population group is not yet ready to practice preventive medicine (i.e., contraception) but is already motivated to effect a cure (i.e., induced abortion). At the highest cultural and socioeconomic level, the birth rate is the lowest as a consequence of the wide use of effective contraception. The incidence of induced abortion now drops precipitously because, at this stage, it is required only as a back-up procedure in those cases where the contraceptive failed or was not used.

Parenthetically, it is worth mentioning that contraceptives are most likely to be used effectively if their administration is completely separated from coitus, and this explains the dramatic acceptance of oral contraceptives and intrauterine devices (IUDs) as well as their high "use-

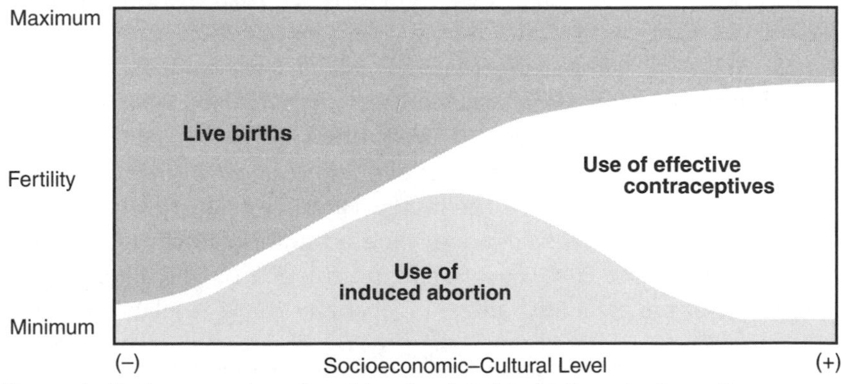

Figure 1. Socioeconomic–cultural levels related to birth control practice.

effectiveness" as contrasted to "theoretical effectiveness." Although induced abortion does not fall within the definition of contraception, it shares, as a fertility control procedure, with oral contraceptives, IUDs, and sterilization the property of being completely separated from sexual intercourse.

ABORTION AS BIRTH CONTROL

It is not generally realized that liberalized abortion laws are of very recent origin and that the first significant legal moves (Scandinavia) did not occur until the 1930s. Nevertheless, once medically supervised abortion is readily available and legally permitted, it can be a powerful method of fertility control on a national scale. In fact, until now, it has been the most effective means in those countries where the proper mix of social and cultural factors (*see* Figure 1) is superimposed on an extensive medical or paramedical establishment. Japan is a classic case in which induced abortion was permitted lawfully only after the last World War and where the precipitous fall in the birth rate coincided with the enormous increase in abortions.

The removal of legal restrictions to induced abortion in Eastern Europe is of even more recent origin (middle 1950s) and has resulted in a very short time in the situation that the abortion rates in four Eastern European countries far exceed that of Japan. Although abortion per se may not be the single most important factor in fertility control, it clearly plays a very significant role[1]—especially in those countries in which there exists already a disposition toward reduction in birth rates, and it behooves us to examine what techniques are currently used and what improvements may possibly be instituted.

Abortion is much safer when performed during the first three months of pregnancy and when it is legally sanctioned. Numerous studies have shown that the mortality rate accompanying abortion is lowest in those countries in which abortion is freely available. For early abortions, the most widely used procedures are either dilation of the uterine cervix followed by curettage or vacuum aspiration. After the 14th week of pregnancy, more complicated procedures have to be employed, the most prevalent one being intra-amniotic instillation of a hypertonic solution (20% aqueous sodium chloride or 50% aqueous glucose). One of the great advantages of legal and freely available abortion

[1]According to a 1992 WHO estimate, worldwide there are performed—daily—150,000 induced abortions—50,000 of them illegal and leading to 500 deaths—every 24 hours.

is that it will encourage the woman who chooses abortion to subject herself to such intervention as soon as possible and with the least amount of delay and danger.

As far as abortion is concerned, until the early 1970s the United States was one of the underdeveloped countries. Even now we are suffering from a restrictive climate that is hardly conducive to a major research effort on chemical abortifacients and related agents. The highly touted "Family Planning Services and Population Research Act of 1970" signed by President Nixon contains two clauses, which are germane to my presentation and which I quote in full:

"Sec. 1004 (a). In order to promote research in the biomedical, contraceptive development, behavioral, and program implementation fields related to family planning and population, the Secretary is authorized to make grants to public or nonprofit private entities and to enter into contracts with public or private entities and individuals for projects for research and research training in such fields."

"Sec. 1008. None of the funds appropriated under this title shall be used in programs *where abortion is a method of family planning*" (italics mine).

CHEMICAL ABORTIFACIENTS

The development of chemical abortifacients—preferably administered by the oral or intravaginal routes—should get a very high priority. However, the development of such agents immediately raises the level of required technical sophistication as compared to simple surgical procedures. This is so because chemical, biochemical, toxicological, physiological, pharmacological, and clinical expertise is required, coupled with the fact that requirements of regulatory bodies (e.g., the U.S. Food and Drug Administration [FDA], the Food and Drug Directorate in Canada, and the Scowen Committee in Great Britain) will now have to be met. In addition, in some countries—notably Japan—there is a built-in opposition of medical personnel involved in surgical abortions who see parts or all of a significant income escaping them.

If we are willing to ignore religious objections—an enormous "if"—then research on chemical abortifacients should be at, or near, the top of the priority scale for future fertility control agents,[2] given the wide (>50 million annually) use of abortion worldwide. Fortunately, at least two

[2]These words were written several years before the research on the chemical abortifacient RU-486 was initiated in France.

important federal funding agencies (Agency for International Development and the Center for Population Research of the National Institutes of Health) seem to have found ways of circumventing the Family Planning Act[3] and have continued to inject significant, though insufficient, financial support into this highly important area of research.

For the sake of convenience, chemical approaches to abortion can be divided into two main classes—namely, agents interfering with embryonic development and agents (e.g., prostaglandins) leading to expulsion of the embryo or fetus. Research dealing with the development of new agents interfering with embryonic development is probably the most difficult from a practical standpoint. Although a suitable agent might be discovered accidentally in connection with other work (e.g., development of cytotoxic drugs), a concerted effort designed specifically for such embryo-toxic agents would have to be based on an extensive, expensive, and carefully designed screening program in subhuman primates.

This leads directly to one of the most serious of all bottlenecks in reproductive biology research—the qualitative and quantitative deficiencies in primate centers. By qualitative deficiency, I mean the relative unavailability of higher primates, such as chimpanzees, and the paucity of biochemical and metabolic data with such animals. Studies on many monkey species have demonstrated surprising differences in their metabolic handling of steroids; furthermore, a WHO scientific group pointed out in 1969 that the most commonly employed primate, the rhesus monkey, differs from humans in its timing and mode of implantation, placental production of chorionic gonadotropin, and in the production and metabolism of estrogenic hormones. A major international effort is required in establishing one or more primate centers that concentrate on breeding facilities for higher primates such as chimpanzees and on extensive reproductive biological studies in order to provide an adequate model for extrapolation to humans. A proposal dealing with the organization of such an international center in the Congo—the world's major reserve of higher primates—has already been made by me in another context.[4]

In view of these difficulties, it seems surprising how many potential agents have already been subjected to some preliminary clinical scru-

[3] This was true during the Nixon–Ford–Carter administrations, but not the 12-year period under Presidents Reagan and Bush.

[4] A detailed description of my efforts to establish a pygmy chimpanzee (*Pan paniscus*) breeding colony in Zaire can be found in my autobiography (*The Pill, Pygmy Chimps, and Degas' Horse,* Basic Books, New York, 1993, paperback).

tiny in humans. However, virtually all of these candidates are antimetabolites, alkylating agents, or other cytotoxic drugs that were administered to limited numbers of pregnant women who also happened to be suffering from serious diseases (usually cancer). The number of cases is far too small to lead to firm conclusions, especially because not all aborted fetuses were examined for malformations; still, some preliminary conclusions can be reached. In the majority of cases, the agent was administered during the first trimester of pregnancy and any fetal malformations that were noted were encountered only in that group. When the cytotoxic agent was given in the second and third trimester, very few abortions were encountered and no malformed babies were delivered.

The abortive efficiency of an embryo-toxic agent would have to be exceedingly high—probably in excess of 99%—before teratological side effects could start to be neglected. In any event, human clinical experiments would have to be backed up by surgical abortions, especially during the early research stages when the minimum effective dose is being determined and method failure is likely to be particularly high.

The likelihood of success in developing an agent that interferes with implantation of the fertilized egg, rather than with the later stages of embryonic development, is considerably higher. Estrogens, notably synthetic ones such as diethylstilbestrol, have been employed with some success as postcoital agents. Although they would not be of any particular use for fertility control in developing countries, in highly advanced countries—where contraception is practiced widely—estrogens could be of utility as an addendum to the contraceptive armamentarium of women. They could be employed postcoitally ("morning-after" Pill) by a woman who had forgotten to use one of the conventional contraceptives or by the person who indulges only in infrequent intercourse. Unless an anti-implantation agent with a substantial safety margin is discovered, a postcoital pill is unlikely to be approved.[5] As noted already in Chapter 5, a once-a-month pill serving as a menses-inducer is technically conceivable. This would be one of the most significant breakthroughs in practical fertility control in underdeveloped as well as highly developed countries. Problems associated with the development of such an agent, however, are nonchemical in nature; aside from political barriers, many requirements of regulatory bodies (notably the FDA)

[5]Administration of high doses of conventional oral contraceptives or ingestion of an antiprogestin, such as RU-486, within 72 hours of unprotected intercourse is highly effective as a morning-after Pill.

are based on highly dubious scientific reasoning to which there is no appeal. The lead times for converting a laboratory discovery into practical reality in the human fertility control field can be enormously long: A decade is a conservative estimate.

We can examine that lead in the abortifacient field—the prostaglandins—which has received the widest scientific attention and (unfortunately) also the most sensational press coverage. Few of the currently available prostaglandins are orally active, and initial clinical experiments have been performed by intravenous infusions as well as intrauterine and intravaginal routes. Of the three, the intravaginal route is potentially the most interesting, because it is the only one, other than oral ingestion, that might lend itself to self-administration by the patient through the use of tampons or other convenient formulations. The largest clinical experience and greatest success rate (up to 90%) has been encountered by the infusion route, notably with patients in the first trimester of pregnancy. However, serious questions have been raised about whether such a method should be used practically during the first 12 weeks of pregnancy, because the presently employed suction curettage is quite safe, requires only a few minutes, and can be carried out on an out-patient basis.

The potential practical utility of prostaglandins in the human fertility control field falls into three main categories. The first is an adjunct to medically supervised abortion and labor induction, and the requirement for medical facilities automatically limits its utility as a widely applicable method of fertility control in underdeveloped countries. However, this application probably can be implemented sooner than any other, and because, by definition, it will be employed largely in more highly developed countries, it is less likely that serious side effects will escape early notice. It would be criminal folly to expose such an agent to too early general and wide use, because the unanticipated occurrence of some side effects, which may well be avoidable during a more carefully designed developmental phase, may totally and permanently wreck the acceptability of prostaglandins in fertility control.

The second practical use would be one that would be based on the self-administration of the agent by the woman without necessary medical supervision or, at most, with advice by paramedical personnel. Only when this can be accomplished do we have at our disposal a powerful fertility control agent that is equally applicable in developed and underdeveloped areas of the world. An orally effective form would be ideal, but in its present absence, the other alternative is to concentrate on other routes that would lend themselves to self-administration: of these the intravaginal or possibly rectal routes appear to be the most promising ones. Even if the problems of formulation, stability, and mode of administration are solved, extensive clinical trials over a period of several years under diverse field conditions are required before such a method can be considered an acceptable one for wide public use. Nothing could kill such a promising approach more effectively than the appearance of a low-incidence side effect (e.g., malformed offspring in case of method failure) that was not anticipated. If all of these requirements are met, then such a prostaglandin abortifacient might be used routinely by a woman if she had missed her menstrual period by a few weeks.

The third potential use is as a menses-inducer but, as yet, there exists an insufficient amount of clinical information to determine how reliable or appropriate such an application will prove to be and what side effects are associated with it. For instance, even a fairly high incidence of nausea or diarrhea might be tolerated by a woman in connection with the very occasional use of a chemical abortifacient, but it is highly unlikely that a large number of women would accept such side effects monthly. Obviously, an agent effective as a menses inducer and as an abortifacient would be exceedingly useful.

OVULATION INHIBITORS AND OTHER HORMONAL AGENTS

The administration of combined progestogen–estrogen preparations (i.e., the conventional "Pill") produces a state that in oversimplified terms may be defined as pseudo-pregnancy. An alternative approach, which mimics more precisely the natural menstrual cycle, is represented by the "sequential" oral contraceptives in which a pure estrogen is administered for 11–15 days followed by 5–10 days of a mixture of estrogen and progestogen. The third and, in many respects, most exciting development is the so-called mini-pill approach. Initial work in this area showed that by the continuous, daily administration of a pure progestational agent in low dose without a mixture with an estrogen (responsible for many of the low-incidence side effects of conventional oral contraceptives), ovulation was not inhibited, menstrual flow was relatively unaffected, and yet pregnancy protection was provided. The mechanism of action is not completely understood but may, in part, be associated with a change in the character of the cervical secretions, which thus provide a hostile environment to the sperm.

Such a continuous low-dose regimen with a pure progestational agent has several advantages that make it particularly desirable for application to women in less developed countries. Indeed, the mini-pill had been marketed in several countries outside the United States and had been used in at least 2 million cycles when a panic response by the FDA resulted in the sudden suspension of clinical work in the United States and the consequent removal of the drug from all world markets. The term "panic response" is used advisedly because it was based exclusively on the ill-advised use (required by the FDA) of beagle dogs for long-term toxicity studies. In spite of the fact that neither monkeys nor humans, exposed to longer periods of time, showed any of these effects, the results from female beagles with their completely different heat cycle (every six months rather than monthly) and notorious sensitivity to steroids was sufficient to cause demands for the instantaneous discontinuance of the drug. No appeal to such decisions is possible, so it is unlikely that such an agent can be resurrected, and similar episodes occurred a few months later with two sequential oral contraceptives. In each instance, the agent had been employed for several years in thousands of women without deleterious effects but was withdrawn because of suspected toxicity in beagles. The reason I am citing these examples is only to illustrate the many quasi-scientific and nonscientific factors that must be taken into consideration in the development of fertility control agents. As a result, major new scientific developments probably cannot be anticipated in the area of steroidal ovulation inhibitors or low-dose progestogens, although this does not preclude significant commercial successes in the markets of highly developed countries through chemically trivial modifications.

Theoretically, one can envisage fertility control approaches for more sophisticated (and sexually disciplined) women through carefully timed ovulation promoted by administration of the hypothalamic releasing hormone—in other words, a chemically induced rhythm method. The difficulties to be overcome in developing such an agent should not be underestimated. Notable problems are the very rapid effect and very short duration of action of this hormone, which make accurate timing difficult. Nevertheless, it is certain that important fundamental advances in human reproductive biology, even if they have no foreseeable practical applications, will be forthcoming.

INTRAUTERINE DEVICES

For fertility control in developing countries, intrauterine devices (IUDs) offer the unique advantage of a one-shot administration, and their relative efficacy—so far only exceeded by oral contraceptives—make them powerful weapons in fertility control. (In China, IUDs are the most widely used method of birth control.) IUDs are the example *par excellence* of how fast an agent can be introduced into wide clinical practice in the area of contraception if it falls outside the responsibility of regulatory bodies in the developed countries, but this state of affairs has changed in the context of a very significant IUD improvement: the "copper T"—a thin copper wire of 1.2–2.0-cm length wound around a T-shaped plastic IUD. The use of a T-shaped device is apparently responsible for a much lowered expulsion rate and less bleeding than is encountered with conventional IUDs, and clinical studies have indicated that the addition of copper results in a much lower pregnancy rate.

The mechanism of action of heavy metals in conjunction with IUDs is incompletely understood, and nothing is as yet known about long-term effects. Preliminary studies indicate that the amount of copper released daily from such IUDs is insufficient to raise the usual plasma levels of copper, but that the local copper concentration in the endometrium and cervical mucus is elevated considerably. Because human sperm is sensitive to copper, this may be a possible reason for the increased efficacy of the "copper-T."

If these preliminary conclusions should be supported by the more extensive clinical trials under way in the United States and abroad (notably Chile, where the original discovery about the beneficial effect of copper was made), then this development is likely to have a significant impact in developing countries. The main problem will probably be to overcome the reluctance of past disillusioned IUD users, who will have to be convinced (no mean feat in some of the population groups under consideration!) that the new device represents a major improvement and to offset some of the emotional factors among more affluent groups in which the opposition to IUDs seems to be most pronounced. Furthermore, because these improved IUDs carry a metal, they are now considered by the FDA as falling within its purview, so that there is no doubt that the time interval between clinical studies and eventual public use is likely to be greatly extended.[6]

STERILIZATION

The ultimate answer to fertility control is sterilization. In males, the most common procedure is vasectomy, which does not impair any of

[6]The notorious Dalkon shield IUD, resulting in multi-billion dollar liability suits, was introduced commercially before medical devices were subject to FDA approval.

his sexual function except for the absence of sperm in his ejaculate. The largest number of vasectomies has been performed in India, but this has little impact on that country's population problem because the vast majority of men who have volunteered for vasectomies already have fathered many children, and in any event their percentage is small. Major advances in vasectomy techniques, which would guarantee reversibility upon demand, would constitute a very significant forward step, although it is likely that its impact would be felt more in populations on the upper end of the socioeconomic–cultural abscissa of Figure 1.

The other alternative, and one that for logistic reasons would be limited to the more highly advanced countries, would be to develop adequate methods for the very long term (10–20 years) storage of human sperm so that every candidate for vasectomy could have sperm preserved, which could be employed in artificial insemination if the vasectomy could not be reversed effectively. An indispensable component of such an approach would be the unequivocal demonstration—so far unavailable—that sperm stored for periods of 20 or more years is still viable and effective in artificial insemination.

Reversible sterility in the male produced by the administration of chemicals falls within the definition of a male contraceptive. However, in Chile, sterilization in females through occlusion of their fallopian tubes has been produced by the intrauterine instillation of quinacrine. The choice of this agent was based on earlier studies in rodents in which various cytotoxic agents were examined for their ability to cause tubal occlusion. These experiments also showed that the process could be reversed by the administration of estrogens or progestogens.

Although the total number of patients was small, the preliminary results indicated that one instillation was insufficient, but that two and preferably three instillations produced tubal occlusion. Out of 12 women who had been given three administrations of quinacrine, only one became pregnant over a period of two and one-half years and the menstrual patterns remained unchanged. Attempts are now under way to determine whether these effects of quinacrine can be reversed by the administration of steroids. Certainly this approach appears very promising and merits extensive follow-up. It does raise a peripheral, but interesting, question. If these clinical studies had been planned in the United States rather than in Chile, would the FDA have approved the clinical protocols and could that much clinical information have been collected in so short a period of time?

MALE CONTRACEPTIVES

If one uses the narrow definition of male contraception in terms of agents employed by the male (as compared to chemical agents affecting sperm, but administered to the female), then the prospects for increased fertility control in this decade through improvements in male contraceptive technology are dim. One can hardly expect substantive advances in the two types of reversible contraception practiced by males—the condom and coitus interruptus—and the likelihood of developing a male Pill in the near future is exceedingly slight.

Even if such an agent were eventually discovered, there remains the question of its acceptability by large segments of the male population, notably in many developing countries. To my knowledge no meaningful surveys have been carried out to determine the attitude of males in various parts of the world toward such antifertility agents, but the psychological and cultural factors would appear to be formidable. For instance, in Latin America the concept of "machismo" and the preoccupation with potency would make it very unlikely that such a male Pill would find significant acceptance unless it could be claimed (and preferably also demonstrated) that it also improved sexual performance. One does not have to be a hypercautious government bureaucrat to realize how difficult it would be to screen for and demonstrate efficacy of a libido-enhancing drug, even though no dearth of male volunteers needs to be anticipated for this type of clinical work.

I conclude that significant developments in male contraception are more likely to occur in those areas in which the agent acts in some manner on the sperm, but where its mode of administration involves the female. The simplest and most harmless of such types is clearly a spermicide employed by a woman prior to intercourse. Although it suffers from the great disadvantage that its use is not divorced from coitus and that it thus requires a conscious precoital act, which clearly results in lower "use-effectiveness," an improved spermicide would, nevertheless, represent an advance in fertility control that could presumably be developed and implemented much faster than more sophisticated systemic methods and would have the additional advantage of possible protection against sexually transmitted diseases. It is remarkable how little chemical talent, ingenuity, and manpower has been dedicated to this area of research. At present, the relatively high pregnancy rate mitigates against currently used formulations, but a tenfold reduction in the pregnancy rate could constitute an important practical achievement in birth control.

A variety of more sophisticated methods can be envisaged that could interfere with the viability or passage of the sperm once it enters the vagina. Let me consider only one, because it is also of great interest from a scientific viewpoint and illustrates some of the deficiencies in human reproductive biology knowledge that still need to be overcome. I am referring to the phenomenon of sperm capacitation, which has been known for nearly 30 years in certain animal species, although it is still not clear whether this effect operates in humans. Thus it has been observed that animal spermatozoa (e.g., rats and rabbits) must reside in the female reproductive tract for a certain period before they gain the ability to fertilize the female's ovum. Furthermore, the seminal plasma of the male's reproductive tract contains a "decapacitation factor" that reverses the "capacitation" phenomenon and makes fertilization impossible. If it can be demonstrated that such factors are also crucial to human fertilization, then *in theory* it ought to be possible to devise an antifertility agent which, when administered locally to the female, would render the male's sperm infertile. There are other alternative approaches that could be developed from such a lead, but they all have one property in common: It is highly unlikely that any would result in practical fertility control agents before the end of this decade.

EPILOGUE

The main purpose of this chapter is to demonstrate that the problem of fertility control is urgent and that the scientific community does not lack theoretical leads, but that the gap between a laboratory discovery and its conversion into a practical agent is widening at a terrifying rate. In the past, scientists from industrial (i.e., pharmaceutical) laboratories have made major contributions that, in fact, have led to most of the new practical contraceptive advances (e.g., steroid oral contraceptives).

One aspect of the scientific retrenchment to which I referred earlier is the continuing estrangement of academic and government scientists from those who work in industrial laboratories. This is particularly noticeable in the area of human fertility control, which has been "discovered" fairly late by government agencies. Planning sessions for future efforts and developments as well as retrospective evaluation of past achievements are now being organized by national and international agencies. A conspicuous feature of these long-overdue activities is the virtually total exclusion of industrial scientists, who, however, are primarily the experts in matters of operational and implementational concern. This fear of "contamination" by industry is especially noteworthy in agencies such as the National Institutes of Health and the World Health Organization, both of whom are now actively involved in fertility control activities. Major symposia on topics such as the metabolic effects of contraceptive steroids—an area where industrial scientists have probably contributed more than anybody else—have been held with total exclusion of scientists from industry. This type of fragmentation of the limited manpower and resources available for serious work on practical developments in fertility control will take the desired goal farther from realization.

7 Male Contraception

In order to understand the vulnerable links in the male reproductive process that might offer novel approaches to male birth control, we need to have at least a simplified view of the complex sequence of events that governs the 82-day life cycle (74 days for spermatogenesis and 7–8 days for maturation) of a human sperm and its habitat before the sperm enters the female in its search for an unfertilized egg.

THE MALE REPRODUCTIVE PROCESS

Spermatogenesis, or sperm production, occurs in the seminiferous tubules contained within the testes. These tubules are more than 700 feet long and produce in excess of 30 million sperm per day. Once formed, the sperm passes into the epididymis, a 20-foot-long duct, where the sperm matures and is stored. During this maturation process, the sperm acquires its fertilizing capacity as well as its ability for independent movement. Subsequently, the sperm is stored and transported in the vas deferens, a duct nearly 1 foot long where the sperm is diluted by the secretions of the seminal vesicle and the prostate to comprise the seminal fluid that is eventually ejaculated through the urethra in the penis into the female genital tract.

Also in the testes are the Leydig cells, which produce the male sex hormone *testosterone*—responsible for the male's secondary sexual char-

Note: Original text written in 1979.

acteristics as well as his libido. Testosterone production is governed by complex hormonal mechanisms. The hypothalamic gland is stimulated by the central nervous system to secrete several hormones called the releasing factors or releasing hormones (RH). The key hormone *regulating male as well as female reproduction* is the "luteinizing hormone-releasing hormone" (LH-RH). This hypothalamic hormone, in turn, stimulates the anterior pituitary gland to secrete two hormones, known as gonadotropins, which are usually referred to simply as LH (luteinizing hormone) and FSH (follicle-stimulating hormone).

In the male, LH stimulates the Leydig cells in the testes to produce testosterone. A complex negative feedback mechanism provides that if too much testosterone is produced, the concentration of LH drops so as not to overload the body with testosterone; conversely, if the concentration of testosterone falls below a certain level, then the concentration of LH rises. The biological role of FSH is much less obvious in the male than in the female. In the male, FSH apparently exerts its effect upon the seminiferous tubules in the testes to produce a polypeptide, inhibin, to which testosterone is bound. Through another negative feedback mechanism, inhibin can lower the circulating levels of FSH if they get too high.

In contrast to the female, who is born with her life's supply of eggs, the male continually produces sperm and thus offers an almost lifelong target for potential genetic damage to the next generation. This condition is important for contraceptive strategy in that chemical interference with the early stages of sperm production is accompanied by the greatest genetic risk, whereas manipulations of later stages in sperm maturation are potentially less dangerous. The most obvious contraceptive strategies in the order of increasing risk are interference with the transport of the sperm before it reaches the seminal vesicle and prostate gland so that ejaculated seminal fluid will not contain sperm; interference with the processes occurring during maturation and storage of the sperm in the epididymis; interference with sperm production in the testes; and interference with the hormonal mechanisms at the anterior pituitary or hypothalamic levels, which would disturb both testosterone production and the production of sperm.

INTERFERENCE WITH HORMONAL BALANCE

More work has been done on interference with the hormonal balance at the pituitary and testicular levels than on any other potential approach to male contraception because, just as with work leading to the female oral contraceptive pill, there was available a body of fundamental knowledge about the hormonal control mechanisms that offered specific points of attack in the male's reproductive process. Unfortunately, hormonal interference in the male has exactly the same inherent disadvantages as it does in the female: It tampers with a very complex interplay of various hormones and thus is likely to lead to a variety of actual or potential undesirable side effects. Moreover, given the present regulatory climate, society's attitude toward safety and risk taking, and what we have learned about potential side effects from 20 years of clinical experience in the female, the development of hormonal contraceptives for males is now likely to take very much longer.

Nevertheless, there is an interesting operational contradiction: Although it will take a very long time (perhaps 12 to 18 years) to bring a male hormonal contraceptive pill to the public, initial clinical experiments with humans can actually be conducted much earlier than might ordinarily be expected. Let me explain. If one followed the traditional path of starting with the chemical synthesis of a *new* substance, one would not reach the stage of small-scale phase I clinical trials for determining toxicity in humans until a minimum of five to six years had elapsed. However, most of the experiments with male hormonal contraceptives have all *started* with such clinical trials because the active ingredient in this instance was not a new substance but the natural hormone testosterone that had been used earlier for other therapeutic applications, either in the male or the female. Thus, initial animal toxicology was not required because major acute toxic effects in humans are not expected in short-term testosterone therapy.

The theory behind attempts to develop a male hormonal contraceptive is relatively simple, but the actual development is complicated. Briefly, what most clinicians have tried to do—in many cases successfully—is inhibit the secretion of the gonadotropic hormone LH through the administration of steroid hormones. Because of the negative feedback loop between testosterone and LH, the administration of high doses of testosterone actually *inhibits* spermatogenesis because the circulating testosterone inhibits the secretion of LH which, in turn, inhibits production of testosterone *within* the testes; this endogenous testosterone is needed for spermatogenesis. The exogenously administered circulating testosterone, however, is sufficient to maintain libido and other secondary sexual characteristics.

Such experiments have actually been conducted in humans, primarily with injectable testosterone preparations because orally effective androgens (male sex hormones) cause substantial liver function impairment if given for long periods of time. Although these preliminary experiments have shown that it is possible to lower the sperm count to levels that lead frequently, though not invariably, to infertility (it is often forgotten that an abnormally low sperm count is no absolute guarantee of infertility), the dosages required are so high that there are very serious questions whether such levels of testosterone can be given for long periods of time to males without danger of major side effects—including prostatic cancer and a variety of cardiovascular and metabolic disorders. Females would incur equivalent risks if their method of contraception were based on the continuous administration of high dosages of estrogen.

Testosterone, however, is not the only steroid that inhibits LH secretion. The two types of female sex hormones, the estrogens and the progestational hormones (both natural progesterone or the synthetic progestational agents present in current female oral contraceptives) will also do the job. Administration of both types of female hormones in men has been studied, but in order to offset the total elimination of the male sex drive, testosterone (either by long-lasting injection or implants) has had to be given concurrently. In principle these methods work, but so far no really satisfactory combination regimen has been discovered that overcomes one of the most serious side effects of such therapy, namely gynaecomastia (breast growth).

INTERFERENCE WITH SPERM TRANSPORT

The simplest way of interfering with the transport of mature sperm prior to ejaculation is to rupture or block the vas deferens. This is accomplished by vasectomy, which involves a simple incision into the scrotum, severance of the vas deferens, tying off of the two ends, and closing of the scrotal incision with a few stitches. The procedure is performed under a local anesthetic and can be completed in less than 30 minutes.

Other than coitus interruptus and the condom, vasectomy is the only method of fertility control practiced by males. Unfortunately, because reversibility is so difficult, vasectomy is for all practical purposes considered only by fathers or middle-aged and post-middle-aged men who are not interested in having more children.

Except for occasional minor short-term side effects shortly following the operation, the longer term effects of vasectomy appear minimal. A possible long-term effect meriting concern is that in an appreciable number of men, reabsorption by the body of the sperm, which after vasectomy does not get removed through ejaculation, may lead to the production of circulating antibodies specific to sperm. The potential consequences of this immunological reaction in the *infertile* man seem to be low, but the re-establishment of fertility in the presence of sperm-specific antibodies is a major question.

PSYCHOLOGICAL EFFECTS OF MALE CONTRACEPTION

Loss of libido is likely to become a real problem in any hormonal contraceptive method, in spite of assurances to the contrary, because of the enormous psychological component associated with sexual potency in the male. The best approach to this problem would be the combination of a male contraceptive with a libido enhancer. Indeed, a libido enhancer per se would be an important contribution, not only commercially but also in terms of alleviating a fair amount of human misery in males. Unfortunately, developing such an agent and, especially, establishing double-blind clinical protocols (in which neither the subject nor the researcher knows the identity of the experimental drug or the placebo) that would lead to FDA clearance for initial clinical research, let alone final marketing approval, is quite another matter.

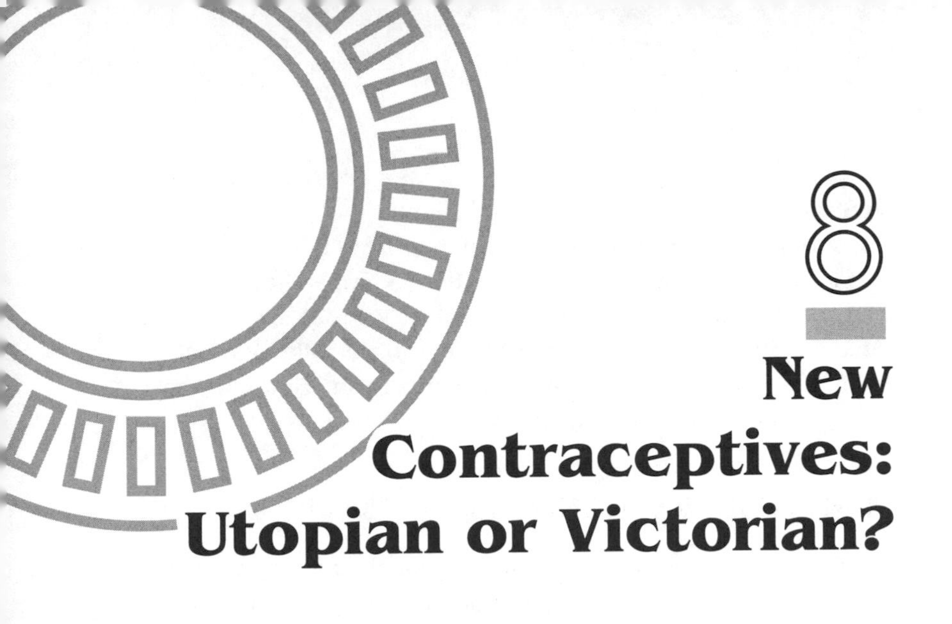

8

New Contraceptives: Utopian or Victorian?

In a letter printed in *The Times,* of London on July 30, 1990, the Principal of the University of Strathclyde wrote that: "among the many issues that confront the world, the most important by far is overpopulation. . . . Most of our energy, environmental, and economic problems and many of our political and social problems are population-driven." Concerns about overpopulation that were first raised in the early 1960s later diminished because of the well-publicized gradual drop in the world's rate of natural increase. Despite this drop, however, the world's population is expected to double by the year 2030 (reaching more than 10 billion). Such an increase is spread across the world no more evenly than are economic or natural resources.

In 1992 two-thirds of the world's population was centered in only 12 countries (Table 1), with Germany (at 81 million) at the bottom of that list as the sole European representative. Assuming 1992 natural growth rates, what will the numbers be 25 years later, an insignificant time interval on the scale of human existence?

Long before 2017, Germany will have disappeared from that list, because its population growth rate is now negative! Newcomers such as Egypt, whose 1992 population of 58 million will double in 30 years, will replace Europe's only representative. Nigeria, the world's tenth largest country in 1992, will double its population in only 23 years, which

Note: Original text written in 1991.

means that Nigeria's population by 2017 may approach that of the United States (the fourth most populated country in the world and also ranked as fourth in terms of land surface). Nigeria is not only incomparably poorer, but it occupies the 31st position in terms of area.

One could cite other horror scenarios, such as Pakistan's; given its current doubling time, it could also approach the United States' current population by the year 2017. Yet Pakistan, potentially an economic "basket case," ranks 35th in terms of land surface.

UTOPIAN POPULATION CONTROL *A LA* 1984

The letter in *The Times* was presumably prompted by such considerations. As a solution, it outlined a form of draconian population control ("approach the problem of contraception in a fundamentally new way: If, for example, contraceptive material were introduced into the human food chain, then all casual sexual encounters would be unproductive"), the naíveté of which demands debunking. Aside from any ethical and political problems, the purely technical ones are of such complexity that they reduce the concept to an absurdity. Twenty years ago, I rebutted

Table 1. The World's 12 Largest Countries in 1992

Country	Population (million)	Rank	PDT[a] Rank (years) at 1990 rate	Rank (surface area)
China	1167	1	53	3
India	882	2	34	7
Former USSR*	284	3	104*	1
USA*	256	4	89*	4
Indonesia	185	5	40	15
Brazil	151	6	37	5
Japan*	124	7	217*	<50
Pakistan	122	8	23	35
Bangladesh	111	9	29	<50
Nigeria	90	10	23	31
Mexico	88	11	30	14
Germany*	81	12	—	<50

[a]PDT is population doubling time

*Demographic problems are under control.

SOURCE: Population Bureau Inc., April 1992.

such Orwellian proposals in an essay (*see* Chapter 5) entitled "Birth Control After 1984" by pointing to several unrealizable requirements, two of which are briefly cited as follows:

The contraceptive would have to be active in the male and/or the female, but only in their reproductive years (up to 60 years in males), and active over an enormous dose range, because intake by an adult versus a two-year-old child would be very different.

If taken in food, the material would have to be incorporated by the supplier rather than by the consumer to ensure universal administration. A dissenter could simply eliminate a given food from his or her diet to escape the contraceptive effect, so it would have to be incorporated into universally required food (e.g., salt). Because everyone must drink water, this might be a better vehicle for such an Orwellian agent. But what about the majority of the world's population getting its water from sources other than central water supplies?

The Times letter ends on the following naively utopian note: "For the protection of the individual, as well as for the protection of the earth, I would therefore suggest that one essential ingredient of milled cereals should be a heat-resistant contraceptive but otherwise harmless chemical additive. Pharmaceutical companies please note."

WITHDRAWAL OF PHARMACEUTICAL INDUSTRY

This last sentence offers a bridge to the realities of this century. If we do indeed wish for new contraceptives, then the participation of the pharmaceutical industry is essential; it is foremost in production and distribution, but also in development, and frequently even in research.

Any pharmaceutical company following the advice of *The Times* letter would, financially speaking, lose its shirt. But even much more modest proposals—say a fundamentally new spermicide or some scientifically feasible though much more ambitious targets, such as a once-a-month menses-inducer or a contraceptive vaccine—have not aroused the interest of any major pharmaceutical company in the United Kingdom or elsewhere. According to a worldwide survey of leading research and development (R&D) therapeutic categories, done by a British firm in 1988, contraceptives did not even appear among the first 35 rankings!

In my earlier essay (Chapter 5), I estimated that the time required to bring two fundamentally new birth control approaches to the market—a once-a-month menses inducer for women and a male pill—is 12–20 years. These long development times, caused by society's understandable risk aversion when dealing with drugs consumed by healthy individuals over periods of years, represent a much more formidable hurdle than the huge financial costs. For instance, what persuasive answer would a man need to the question, "Will I get cancer or (even worse) become prematurely impotent, if I take my pill for the next 40 years?" before a male contraceptive pill could be marketed?

THE CURRENT SITUATION

Deliberate and reversible birth control is effected primarily through steroid contraceptives, condoms and other (less widely used) barrier methods (diaphragm, sponge, and cervical cap), intrauterine devices (IUDs), abortion, coitus interruptus, and determination of the "safe" period. Sterilization, though theoretically reversible in males and females, is best considered an irreversible procedure. The use of these methods differs widely because of regulatory, religious, cultural, and economic factors. One need only consider the differences in contraceptive practice among Japan, which still lacks regulatory approval of the Pill; the United States, which has virtually discontinued the use of IUDs; and China, where IUDs are the method of choice.

The demographic dimensions of the current economic problems of Third World countries are so serious that most are taking at least modest, if not heroic, steps to reduce population growth. The urgency of that problem means that they cannot wait for the development of new methods. But what about the affluent countries? Most have their demographic problems under control (*see* the asterisked entries in Table 1), although the quality of birth control, countrywide or in individual subsets of their respective populations, may well be poor. These are also the countries where most technical manpower and resources reside, so let us examine the rationale for the kind of improvements in birth control that would make the most sense from a policy standpoint and then determine what the likelihood is of achieving such goals.

Of the technologically advanced and industrialized countries, only Russia, the United States, and perhaps Japan will remain on the list (Table 1) of the 12 most populated countries for the next couple of decades. Instead of being technological leaders in an effort to extend the range and quality of human birth control, all of them suffer from pronounced stagnation. Yet all three countries suffer from acute myopia in not responding to the stark fact that the incidence of abortion reflects the state of contraception. Russia has the highest abortion rate in the world as a consequence of its deplorable state of contraception. Japan, though racing at the technological and economic level into the 21st century, still depends on three pre-War methods of birth control: condoms, the calendar-based Ogino rhythm method, and abortion. And the United States? It has the highest teenage pregnancy and abortion rate of any industrialized country. Although improved birth control is not the only answer, a wider choice of better methods would clearly help.

There are at least 10 interventions (*see* Table 2) in the human reproductive cycle, divisible into pre- and postcoital approaches, that could result in effective birth control. For obvious biological reasons, postcoital birth control is limited to women. At present, abortion is the sole available postcoital approach (used by approximately 50 million women worldwide). The "morning-after" Pill, consisting of high dosages of orally effective progestins or estrogens ingested within a few days of intercourse, is acceptable only as an occasional emergency step because of the otherwise grossly excessive, cumulative exposure to potent hormones.

For the individual believing that any interference with a fertilized egg—however recent the fertilization—constitutes abortion, postcoital contraception would be acceptable only for a very short time interval: the period during which the sperm travels toward the ovum. However, for millions of women, interference with a fertilized egg for a few days or even three or four weeks following conception would still be equated

Table 2. Steps Susceptible to Regulation

Step	Administered to
Pre-coital	
spermatogenesis	male
sperm maturation	male
sperm motility	female/male
ovum fertilization	female/male
ovulation	female
Post-coital	
luteal function	female
fertilization	female
ovum transport	female
implantation	female
embryonic development	female

with contraception. This huge group clearly would benefit from advances in the more broadly defined area of postcoital contraception.

MENSES INDUCTION

If I were restricted to choosing a single new contraceptive, it would be a once-a-month pill effective as a menses-inducer. Instead of currently used oral contraceptives, which are taken daily for most of the month, a menses-inducer would be ingested by a woman only during those months when she had unprotected coitus. Instead of waiting to see whether she had missed her period, a woman would take a single pill (containing a short-lived and rapidly metabolized drug) to induce menstrual flow at the expected time. Although not acceptable or suitable to every woman, for many such a regimen would represent an enormous improvement: at most, she would be taking 12 pills annually, rather than the present 250 or more. With such a novel pill, women would not know whether they carried a fertilized egg. The single most important feature of such a method is that the decision to contracept is made postcoitally.

In the early 1980s a group of French investigators (Georges Teutsch, Daniel Philibert, and André Ulmann of Roussel-Uclaf, and Etienne-Emile Baulieu of INSERM in Paris) reported that a synthetic steroid progesterone antagonist, named RU-486, possessed such menses-inducing properties. Although not suitable for regular, monthly menses-induction, RU-486 nevertheless has turned out to be the single most important new development of the past two decades in birth control as an important alternative to surgical abortion. A single, oral ingestion of RU-486 (after confirmed pregnancy, but not later than seven weeks after the last menstruation) followed two days later by a single intramuscular, intravaginal, or oral administration of a prostaglandin, resulted in heavy bleeding with complete expulsion of the embryo in 96% of the women who tested it.

By now, at least one-third of all abortions performed in France are based on this RU-486 regimen. In addition to eliminating the cost and burden of surgical intervention and anesthesia, this method also puts a premium on very early abortion: precisely what is recommended on safety grounds to any woman choosing to abort. Considering that worldwide, nearly 200,000 women die annually as a result of botched abortions, alternatives such as RU-486 must be given serious consideration.

PROSPECTS FOR NEW METHODS OF BIRTH CONTROL

Since 1983 the World Health Organization (WHO) has sponsored clinical trials with RU-486 in a variety of countries (China, India, Singapore, Cuba, Italy, Hungary, etc.) to examine its performance in women of different ethnic backgrounds. The United Kingdom, some of the Scandinavian countries, Holland, and China have or will shortly approve use of this drug as an alternative to surgical abortion. Yet in the United States, where abortion is legal and is used annually by 1.6 million women (at least one-quarter of them teenagers), no pharmaceutical company has dared to apply for government approval in the light of a virulent, antiabortion campaign centered on RU-486 and supported vigorously during the presidencies of Ronald Reagan and George Bush.

The American government did not confine its opposition to its national borders. Even though the United States did not contribute to the WHO's "Special Programme of Research, Development, and Research Training in Human Reproduction"—an international effort heavily supported by the Scandinavian countries, the United Kingdom, Germany, Canada, and many other governments—the U.S. State Department in 1991 saw fit to question whether the WHO used World Bank funds for some of its clinical research on RU-486. The implication seemed to be that the United States might reconsider its continued funding of World Bank activities.

I cite these details to illustrate the enormous politicization of birth control in my country, which is one reason why no large U.S. pharmaceutical company is pursuing work on progesterone antagonists in women, male contraception, or contraceptive vaccine developments, to cite only three novel and scientifically feasible approaches to birth control (all of which have been investigated in animals and humans under the umbrella of the WHO program and other agencies). Aside from relatively minor improvements in current birth control methodology, such as improved delivery systems and dosage forms for steroid contraceptives, there seems to be little on the horizon to fill the shelves of a contraceptive supermarket by the turn of the century. The reason is obvious: the politics of contraception, rather than science, now plays the dominant role in shaping the future of that field.

JET-AGE RHYTHM METHOD

Rather than ending on such a discouraging note, let me turn to "natural family planning" (NFP), "rhythm method," "periodic abstinence,"

"Vatican roulette," or whatever euphemistic or pejorative term one wishes to use to describe the determination of the "safe" period. David Lodge, a wise and humorous British novelist had this to say in his book *Souls and Bodies* (Penguin, 1980):

> Clerical and medical apologists . . . encouraged the faithful with assurances that Science would soon make the Safe Method as reliable as artificial contraception. But the greater the efforts made to achieve this goal, the more difficult it became to distinguish between the permitted and forbidden methods. There is nothing, for instance, noticeably "natural" about sticking a thermometer up your rectum every morning compared to slipping a diaphragm into your vagina at night. And if the happy day *does* ever dawn when the Safe Method is pronounced as reliable as the Pill, what possible reason, apart from medical or economic considerations, could there be for choosing one method rather than the other?

Considering that I concur with Lodge's opinion, why have I called for a new look at one of the oldest and, in many respects, least reliable methods of birth control? Strangely, determination of the safe period is one of the few areas of contraception where current scientific advances may actually serve to overcome some of the political obstacles, and do so with some extra bonuses outside the specific realm of birth control. Aside from the inconvenience of daily record-keeping, the "symptothermal" method of NFP (keeping track of changes in basal body temperature as well as in the viscosity of cervical mucus) requires an average of 17 days of abstinence from intravaginal intercourse.

In principle, accurate determination of the onset of ovulation could reduce this period of abstinence by more than 50% and thus improve significantly NFP's poor acceptability and efficacy. Because the fertile period of an egg is short (approximately one day), precise knowledge of the passage of ovulation would provide a "green light": unprotected coitus is now safe. This can be established by a rise, following ovulation (i.e., the post-ovulatory or luteal phase of the menstrual cycle) of the female sex hormone progesterone (or its metabolites) in blood, saliva, or urine. To cover the first half (i.e., pre-ovulatory or follicular phase) of the month, prediction of ovulation by three or four days is needed, which can be accomplished by detecting the rise of the female sex hormone estradiol in a woman's body fluids. Such an advance warning ("red light") is needed, because sperm can remain viable in a woman's fertile mucus for about three days.

During the past two decades, advances in analytical biochemistry (notably radioimmunoassays) have made possible the accurate determination of such hormonal changes in the laboratory. More recently, monoclonal antibody techniques in ingenious formulations have made it possible to detect visually the rise in progesterone metabolites in one drop of urine in less than one minute by the appearance of a colored spot. In other words, a convenient "green light" in the home is now available by means of two, or maximally three, such tests during the luteal phase.

Although the absolute concentrations of estrogenic hormones and metabolites are considerably lower than those of progesterone, there is no question that a similar "red light" home test, based on detecting the pre-ovulatory rise in estrogens, can be developed for urine within two to three years. Is such an effort worthwhile?

FERTILITY AWARENESS

As a means of birth control in Third World countries, such a high-tech method would be useless on financial and operational grounds alone. Given the poor image of NFP, even in advanced countries such as the United States, the initial impact in terms of improved birth control is likely to be low, although any addition to the contraceptive supermarket is to be encouraged. The cost of such a combined red light–green light test, involving approximately five tests per month, would initially equal that of a monthly supply of oral contraceptives—thus making it acceptable to a segment of those women who already practice NFP or do not tolerate the Pill.

However, I would focus on a potentially much wider customer base under the banner of fertility awareness. I am not referring to women with known infertility problems; a somewhat limited, though highly committed group, through whose cooperation many advances in hormonal detection of ovulation have first been accomplished, and for whom neither cost nor inconvenience are disincentives. Rather, I am thinking of the many women who now feel strongly about health awareness and about making more health-related decisions by themselves: for instance, the rapidly increasing number of professional women who postpone childbearing until their late thirties, or the serious female joggers and athletes.

For many a woman in our affluent society, knowing whether and when she is ovulating should be a routine item of personal health information that could have advantages quite separate from birth control. Two major surveys conducted recently at Stanford University have shown that the majority of women would be interested in purchasing and using such a biochemical ovulation kit (named "The Wizard of Ov" by one student), irrespective of sexual activity. Many physicians, especially epidemiologists studying the incidence of cancer in female reproductive organs, might find a long-time record of ovulatory behavior extremely useful. Evidence for excessive calcium loss in women who menstruate regularly, yet do not ovulate, is another example where such knowledge might offer useful information. Why not employ such hormonal methods of ovulation detection and prediction as routine teaching tools in high schools? Emphasis on fertility awareness rather than birth control may be an effective strategy to fight the continuing politicization of sex education in American high schools. Who knows? It may even lead to a reduction in unwanted pregnancies.

OVULATION DETECTION AND POSTCOITAL CONTRACEPTION

Finally, precise and convenient determination of ovulation may be much more than a jet-age rhythm method. It may even lead to a jet-age contraceptive method that could result in a substantial reduction in abortions. In 1991 Marc Bygdeman and colleagues of the Karolinska Institute in Stockholm, who first introduced the use of prostaglandin in conjunction with RU-486, made the intriguing clinical observation that the administration of a single pill of RU-486 two days after ovulation prevented implantation of a fertilized egg without further disruption of the next menstrual flow. Presumably any antiprogestin, not just RU-486, would have the same effect. Although more extensive clinical work is required, the prospect for a major advance in postcoital contraception is becoming brighter, provided it is combined with a convenient and affordable home test for ovulation detection.

For comparatively affluent, educated, and motivated couples, the prognosis for significant advances in birth control is dim, though not hopeless. But the rest of the world will have to depend for a long time on existing methods or minor modifications thereof.

9

Searching for Ideal Contraceptives

Almost every 30 seconds, another teenager in the United States becomes pregnant. Attacking the problem of rampant teenage pregnancies requires major long-term approaches on the educational, social, economic, and political fronts as well as on the purely operational one dealing with contraception. Just working on the last is at best a Band-Aid, although if wounded I would not sneer at Band-Aids in the absence of other treatment. Improved contraception—in terms of inherently improved activity and improved acceptance by teenagers—coupled with the necessary socioeconomic and cultural changes could make a great difference. Can an "ideal" contraceptive for teenagers be defined and created? If yes, when?

Oral contraceptives, condoms, and coitus interruptus are the most commonly used teenage contraceptives; the diaphragm and the "safe" period follow far behind. This is especially true of the younger and poorer group of teenagers who are sexually active. None of these methods has been designed specifically for teenagers, nor are there any specific formulations. What these methods all have in common is that none of them are postcoital. Except for coitus interruptus, all other contraceptives are precoital. Yet, one of the most common methods of birth control practiced by teenagers is postcoital, namely abortion. One-fourth of

Note: Original text written in 1985.

all induced abortions in the United States are performed on women below the age of 19—a record for a Western industrialized country.

Many American teenagers get pregnant in spite of the availability of efficacious contraceptives such as the Pill or condoms. Some of the reasons for this situation deal with availability, others with acceptability. Given that the development of a fundamentally new contraceptive agent cannot be anticipated before the next century, I believe we should first make the use of presently available contraceptives easier and more attractive for teenagers and then create a new teenage contraceptive.

Among the presently available methods, the condom is the safest and, if used properly, is quite efficacious. It has the unique advantage over all other contraceptive methods of preventing the spread of several venereal diseases—an important problem among teenagers. Recent developments in condom manufacture—thinner ones, colored ones, spermicide-containing ones—represent significant improvements, either in terms of acceptability or efficacy. Also, efforts have been made to teach teenage males that they should consider assuming, or at least sharing, the responsibility for contraception.

How can condom use by teenagers be improved? The steps are simple. First, make condom purchases as private and anonymous as possible to eliminate the embarrassment of many teenagers buying his (or, God forbid, her) first condom in a drug store. Dispensers in readily available public sites (such as toilets in most, rather than a few gas stations) would be an example. Second, make condoms for teenagers very cheap. Gasoline station toilets are clearly not just frequented by teenagers. The ideal place would be toilets or other sites in public school gymnasiums. A marvelous economic case could be made for selling such condoms for a dime or less—a societal bargain even if many condoms end up being filled with water rather than semen. The chances of implementing such a program in most American junior and senior high schools is close to zero.[1]

The diaphragm, when used properly, is a reasonably effective and completely safe method of birth control for the motivated and educated woman. In my opinion, neither it nor the cervical cap will ever have a significant impact on reducing teenage pregnancies. The only teenage group currently using diaphragms is precisely that which, for educational and socioeconomic reasons, is the one least likely to have a high

[1]These words were written in 1984, before the dimensions of the AIDS epidemic were realized and when all three major national TV networks had refused to air any birth control messages.

pregnancy rate. Many a teenage virgin is unlikely to attend classes—even if they were available in high schools—that would teach her how to insert a diaphragm. The embarrassment factor in such classes would also be an enormous barrier.

The contraceptive sponge could play a significant role. The descriptive material accompanying the sponge is simple; the device can be bought without a prescription; and the potential users can experiment with its insertion in total privacy. In many respects, it is the ideal barrier method for female teenagers, and it does not suffer from the "messiness" syndrome—the female counterpart to the male's complaints about lack of sensation with condoms. Is there anything wrong with the sponge, and how can it be improved? The problems are essentially the same as with condoms. Just as the young male can experiment (that is, masturbate) with a condom in privacy, a young woman can experiment with the insertion of the sponge. However, she can buy it only in a drugstore, and there are many fourteen- or fifteen-year-old women who would be too embarrassed to do so and, on occasion, even reluctant to ask an older surrogate to purchase one. The current price is too high for the sexually highly active, impoverished user, but this per se is not a reason it will not be acceptable to many teenagers. Many of them have intercourse infrequently, and their family income levels would not present a significant obstacle.

What is missing are promotion and distribution channels specifically designed for teenagers. In terms of distribution, dispensers in accessible sites (for example, women's washrooms, next to tampon dispensers) of sponges at a subsidized price (say, 25 cents per sponge) would be an answer. The likelihood of such action is about as low as it is for the placement of condom dispensers in high school gyms. At present, promotion to teenagers is basically a word-of-mouth proposition, except for the teenager's exposure to the standard magazine advertisements which do not call attention to the sponge's specific features suitable for teenagers. This is clearly better than no action at all, but it is typical of today's unrealistic and moralistic American approach to teenage pregnancy. One promotional feature of the original marketing strategy of the TODAY sponge is relevant to teenagers. The company advertised a 24-hour hot line that women could use anonymously and cost-free to get responses to queries associated with the sponge. If the existence of such a hot line were widely publicized among teenagers, it could have a significant impact.

Why are oral contraceptives the most widely used method among teenagers? It is the ease of administration, the availability through teenage programs at Planned Parenthood centers, and the fact that the woman can take the Pill without anyone's knowledge, especially her male partner's. She does not have to acknowledge that she is prepared for sexual intercourse as she would have to with barrier methods.

Why is the continuation rate so unsatisfactory among teenagers? The various reasons—perceived side effects, the need for a prescription, taking a daily Pill in preparation for only occasional intercourse, and many others—are well known. Some of these objections apply to all women, not just teenagers, and some can be overcome whereas others cannot. For instance, many arguments have been made that the Pill should be sold over the counter with appropriate warnings. An insignificant number of inappropriate Pill users among young women is prevented by the need for prescriptions, and a much larger proportion of appropriate Pill users is prevented from access to the Pill and hence exposed to the much greater risks of teenage pregnancy. Yet taking the Pill off prescription has very little chance in the United States for the balance of this century, even though this has happened de jure or de facto in many countries.

A different evaluation and public dissemination of the lower risks of the Pill among young women and of the noncontraceptive beneficial side effects of the Pill would increase its use among teenagers. Such a selective publicity campaign would be fought by a wide and disparate American constituency, ranging from feminists and consumer groups to religious fundamentalists. In theory, a sustained-release contraceptive like Depo-Provera is almost tailor-made for teenagers in that it does not require daily attention or repeated motivation. Depo-Provera has caused more arguments, clinical and emotional, than any other contraceptive, and there is very little chance that such an agent will be recommended specifically for teenagers. Sustained-release steroid contraceptives such as Norplant have the disadvantage for teenagers of having been designed for long-term duration. Such implants in the upper arm can be removed by a simple surgical procedure, but the psychological impact on the teenager, and especially on the parents, of committing a school-age person to a five-year contraceptive, seems to me significant, at least in the United States.

Among teenagers, primarily because of lack of knowledge or even discipline, depending on the "safe" period of a woman's menstrual cycle is even more of a gamble than it is among the average female population. We can expect significant technical advances through the probable development of a convenient dip-stick method of predicting when ovulation has occurred, but while such dipsticks could be of substantial utility in educational programs dealing with the menstrual cycle, their impact on teenage pregnancy rates is likely to be negligible.

Even if all of the theoretical improvements of existing methods were implemented, a large number of American teenagers would still get pregnant. Why? Numerous surveys have shown that many teenagers do not take recourse to contraception until they have been sexually active for some time and frequently only after they have become pregnant. For this reason, the future "ideal" teenage contraceptive ought to be a postcoital rather than precoital type. It would have an impact on teenage pregnancies as well as abortions.

10

Steroid Contraceptives in the People's Republic of China

The history of population policy and contraception in the People's Republic of China (PRC) is fairly well documented from the period of the initial Communist victory in 1949 until the beginning of the Cultural Revolution in 1966. But for the next six years, reports have been sporadic, anecdotal, and frequently even contradictory. Compare the following two 1971 quotations: "All means of family planning and contraception are available except for the scarcity of hormonal contraceptive pills which are considered too expensive and possibly dangerous to the mother's health." "The present pill is now manufactured in the billions in the largest chain of labs of this nature in the world."

In 1973 I made a lecture trip to the PRC as one of the first American scientists invited to that country following Henry Kissinger's sensational appearance in China. Extensive discussions with and presentations by staff members of the Institute of Materia Medica of the Chinese Academy of Medical Sciences in Beijing and the Institute of Organic Chemistry of the Chinese Academy of Sciences in Shanghai demonstrated quite clearly that the chemical competence in the steroid and prostaglandin fields was roughly comparable to that existing in North America and Europe. Since the time of the Cultural Revolution, scientists in these laboratories had completed the total synthesis of the steroid contraceptive norgestrel and several of the naturally occurring

Note: Original text written in 1973.

prostaglandins. These syntheses were exceedingly complicated and involved a high degree of chemical sophistication and technical infrastructure.

Even more important, substantial expertise in the steroid chemical field already existed in the early 1960s, primarily through the efforts of Huang Minlon, a distinguished steroid chemist trained at Harvard and the industrial laboratories of Schering, A. G. in Berlin. On the basis of early steroid work at these two research institutions, norethindrone (the same substance synthesized by our group at Syntex in Mexico City in 1951) and the estrogen ethynylestradiol were produced industrially, starting with diosgenin from Chinese yams. Initial clinical work with a dose (2.5 mg of norethindrone and 0.5–0.6 mg of ethynylestradiol) commonly used at that time in Western countries was performed in Shanghai. By 1965 the dose of this Chinese "Pill No. 1" was reduced by half and, during the Cultural Revolution, the most popular dose (0.625 mg of norethindrone and 0.035 mg of ethynylestradiol) was developed clinically. By 1969 this Pill had become the oral contraceptive of choice. Only in 1973 did the Food and Drug Administration (FDA) consider similar low-strength combinations of norethindrone for approval in the United States!

Even more remarkable than the high level of chemical capability and relatively early date of Chinese clinical work with oral contraceptives was the volume of manufacture. In 1973, Shanghai Pharmaceutical Factory No. 7 (all package labels were identified by factory and date of manufacture) was producing approximately 2.4 billion tablets per year (low-dose Pill No. 1 together with a small amount of the higher-dose variant and of Pill No. 2, a different combination of hormones)—sufficient for approximately 9.6 million women per year. This factory was not the only one producing oral contraceptives in the PRC; for instance, Shanghai Pharmaceutical Factory No. 11 supplied the oral contraceptives for the Guangzhou region. These two factories carried out the final tableting; other factories performed the actual multistep chemical synthesis starting with diosgenin.

Quality control of these tablets was fairly primitive. I collected samples in various cities from sources such as street pharmacies, commune dispensaries, and pharmaceutical factories for analysis in the United States according to FDA standards. The content uniformity of their pills in a given bottle appeared to be below FDA standards, but the actual quality of the tablets and of "Birth Control Injection No. 1" generally fell well within their specified limits (±15%). As far as contraceptive efficiency was concerned, such limits appeared to be quite satisfactory.

By far the most interesting material encountered was a "paper formulation" referred to as "sheet type oral birth control pill," developed at Shanghai Pharmaceutical Factory No. 7. (I collected material with manufacturing dates of March 25, 1972, and March 14, 1973.) The progestational and estrogenic steroids were deposited on colored water-soluble carboxymethylcellulose paper (each steroid preparation was of a different color), which was first cut into strips and then perforated to yield the "daily square." The "monthly sheet" containing 22 squares together with a package insert (same color as medicated paper) was then packaged automatically in a cellophane envelope. The Chinese technicians cited four advantages of this formulation: greater protection for the workers, who were much less exposed to large quantities of active steroids as compared with tablet manufacture; a much smaller and lighter monthly package, which was much easier to ship and store; a more uniform product; and economic raw materials (no tablets, bottles, etc.) and simpler technology, requiring fewer people and much less equipment and space.

During lectures and discussions in Guangzhou, Shanghai, and especially Beijing, I compared the regulatory and operational problems for chemical contraceptives in the United States with "FDA-type" procedures in the PRC. The Chinese modus operandi appeared to be more flexible and considerably looser than ours. Animal toxicity requirements did not exceed 6–12 months (as compared with requirements of up to 10 years in this country). The decision for clinical testing was made during "discussions" between the laboratory scientists, clinicians, and representatives of the health authorities, and a similar "discussion"—including also representatives of the pharmaceutical factory—was undertaken before broad-scale distribution. The justification for this ad hoc procedure was that the officials wanted to alleviate human suffering as quickly as possible. In their discussions, compromises were supposedly always reached; it was said that "There is no red tape [presumably referring specifically to drug approval] in China because everybody wants to keep safety and health uppermost in mind." I did not pursue the question further in terms of the argument raging in America of how to resolve the dilemma between "as quickly as possible" on the one hand and "safety and health" on the other.

What about the official policy on fertility control in the PRC? Propaganda on family planning was ubiquitous, and the desirability of limiting a family to two children was greatly emphasized. The remarkable increase in tubal ligation of mothers with two to three children (in several urban communities the number of sterilized women is supposed to exceed that of contraceptive users) testifies to the apparent success of such exhortations to limit family size. Dissemination of information on contraceptive practice itself, however, was essentially absent in educational institutions, including middle schools and universities, and was limited to married persons. The virtual absence of premarital intercourse and the great pressure to postpone marriage obviously had a substantial impact on birth rates in urban areas, notably Shanghai and Beijing.

In 1973, I visualized future collaboration in the fertility control field with the PRC in the following fields: basic sciences, clinical (including epidemiologic) studies, demographic studies, sociologic and cultural aspects, and commercial relations. In my opinion, however, there was little likelihood of important interaction in the last three areas. PRC officials would consider extensive inquiries into demography an inappropriate intrusion in their own affairs, as was specifically stated by then Premier Chou En-Lai in response to a question I posed to him at a May 27 interview. Commercial relations in the birth-control field were exceedingly unlikely because of China's total self-sufficiency in contraceptive hardware and the country's potential to manufacture whatever chemical or device might be developed abroad.

However, I believed in the potential worldwide collaboration in the field of clinical testing and especially in epidemiologic studies. The lag time between the initial laboratory discovery and final clinical application was then much shorter in the PRC than in the United States. If there were substantial agreement on the clinical protocols employed, many of the clinical results from the PRC could be of immediate benefit to many other countries of the world. For example, Pill No. 1 had been in existence in the PRC for several years whereas its American counterpart dosage was not introduced until 1973. Surely, I thought, knowledge that millions of Chinese women had been using a given formulation would (or should) have some constructive impact on foreign regulatory agencies, especially if they were also aware of the type of clinical protocol and follow-up procedures employed in the PRC. Non-Chinese specialists could also make useful contributions to their Chinese colleagues on more sophisticated automated clinical laboratory tests that perhaps were not being used in the PRC.

To me, the most important area was in epidemiologic studies. Because they would be of a very long-term nature, it was critical that the protocols and statistical database be established very early. In this field, Chinese contributions could be of global impact, and oral contraceptives could be used as a typical illustration; the PRC probably had more women on oral contraceptive therapy than did any other country. However, in contrast to so many women in North America and Europe, their Chinese counterparts were much less mobile, jobs and residences changed rarely, and local record keeping at the site of job or residence (or both) was potentially unsurpassed. Furthermore, in contrast to North American and European females, Chinese women were not exposed to a plethora of oral contraceptives but rather to only Pill No. 1 (based on norethindrone) or Pill No. 2 (containing megestrol, which was discontinued in the United States in the 1970s).

At the same time, large numbers of women who were living in essentially the same locale and were exposed to identical nutritional and environmental factors had been sterilized or had never been exposed to oral contraceptives because they were using intrauterine devices or other therapy. For the epidemiologist, such cohort studies could approach Nirvana if certain additional conditions were met: agreement about the nature of the follow-up observations, internationally accepted and understood methods of data retrieval, possible compatibility with various computers, and final statistical evaluation.

CODA (1981)

Since my 1973 visit to the PRC, significant progress has been made in the area of birth control. Western contraceptive hardware is widely available and free of charge. Among younger women, oral contraceptives are the leading method (for women of all ages, IUDs are the most widely used form of contraception), with abortion available on demand. Many Chinese men use condoms. Sterilization is heavily promoted by the state; it is claimed that in some of the urban centers nearly half the otherwise fertile couples have been protected through sterilization.

Most significantly, the Chinese have also developed new fertility control methods. For example, a synthetic steroid called anordrin has been marketed as a "vacation pill" to satisfy a peculiarly Chinese requirement for birth control in couples who are separated most of the year and cohabit only on vacation. But the most interesting Chinese contribution is an experimental male contraceptive pill, based on the cottonseed constituent gossypol, which was first announced in late 1978. These Chinese studies, based on clinical studies with thousands of men over a period of several years, have stimulated a great deal of activity in the male contraceptive field, but it remains to be seen whether a clinically useful male anti-fertility agent will materialize from this lead.

Even more important than this new hardware, however, are some of the software issues—cultural and quasi-legal aspects of birth control—that are almost unique to that country. Because premarital sex is strongly discouraged among Chinese, postponing the age of marriage to the middle or late 20s has a major fertility-limiting effect. China is the only country in the world that promotes the one-child family as the ideal through incentives such as the "Planned Parenthood Glory Coupon," which carries with it certain benefits in health care, food allowance, space allocation, and work assignments. If additional children are born the benefits are withdrawn.

Broadcasts from Beijing indicate that some of the disincentives border on the draconian. For instance, a broadcast on April 12, 1980, described what happened to a department head of the Beijing Number 3 ball-bearing factory and his wife, who was employed by the Beijing steel plate factory. Both were party members, but "they ignored the calls and regulations of planned parenthood and had their fourth child in January of this year." As a result they were charged a 15% excess child fee from the period of pregnancy until the child reaches the age of 14; the salaries of both husband and wife were reduced. They had to pay consultation fees during pregnancy, as well as hospital and delivery charges, and they were not entitled to maternity allowances. The wife received no salary at all during the maternity leave. The husband received no bonuses for one year, and the wife received none for three years. In addition, they had to return their year-end bonus for 1979. The total penalty amounted to 3,000 yuan, well over a year's salary.

Even more dramatic was a June 26, 1980, broadcast reporting that more than 100 party and municipal officials in a provincial city underwent sterilization to take the blame for their laxity in enforcing birth control: "When the provincial government criticized them officially they hastily conducted self-criticism and . . . underwent vasectomy or tubal ligation. . . ." I doubt whether any other country in the world is in a position to inspire such reproductive self-criticism. And I am certain that none would be voluntarily inclined to do so.

11

The Bitter Pill

In 1951, when our research group in Mexico City accomplished the first synthesis of an oral contraceptive, Mexico had 28 million inhabitants; in 1989, it is now approaching 90 million and has risen to 11th rank among the most populous countries. Mexico City is one of the largest cities in the world, and by the turn of this century its population will probably equal that of the entire country in the year of the Pill's first synthesis. During that process of explosive growth, it has also become the most polluted city in the world.

The population growth of Mexico after World War II is not unique. In 1923, the year of my birth, the world's population was 1.9 billion. On my 65th birthday, it had exceeded 5 billion, and at the present growth rate it will reach 8 billion on my 100th birthday. Today the populations of Europe and Africa are almost identical. In just 35 years—in spite of famine and disease—Africa's population will triple unless the acquired immunodeficiency syndrome (AIDS) epidemic interferes with conventional demographic predictions. Europe's population will stay essentially the same.

Yet this is not the bitter pill of my title, which refers to the fact that the United States is the only country other than Iran in which the birth

Note: Original text written in 1989.

control clock has been set backward since 1980. The quality of birth control in the United States is not likely to change by the year 2000, with the consequent likelihood that there will be no significant reduction in the number (ca. 1.5 million) of abortions that now take place annually in the United States. Indeed, the contraceptive choices in the United States at the end of this century may be even more limited than they are now.

INTRODUCTION OF STEROID ORAL CONTRACEPTIVES

Until the introduction of the birth control pill in the early 1960s, abortion (then illegal in all but a few countries) was virtually the only available method of birth control separated from coitus. In my opinion, this property of the Pill (its separation from coitus) and the privacy it offered a woman, rather than the Pill's efficacy, made its initial acceptance so rapid. By the end of that decade, the cumulative decisions of nearly 10 million American women had made the Pill the most popular method of birth control.

Yet even now, three decades later, epidemiological reports regularly include the debate whether prolonged use of the Pill increases the risk of breast cancer. Many women will be discouraged to learn that a conclusive answer will not be available before the turn of this century, because the dosages of both the progestational and the estrogenic components of the Pill have been progressively lowered since the middle-1970s. The same reservation also applies to some of the beneficial, noncontraceptive effects of the Pill, such as protection from benign breast tumors and ovarian, as well as endometrial, cancers. Will the protection persist for women taking the lower dosage steroid regimens?

The introduction of the Pill into medicine came at the best possible time, and also at the worst. It was a time, before the thalidomide tragedy, when new drugs were rapidly being introduced; pharmaceutical companies, the media, and the public proclaimed and accepted the benefits of the postwar chemotherapeutic revolution. Every problem, be it a medical one or a social one such as the population explosion, seemed amenable to a "technological fix." It also proved to be the worst of times, because the same decade saw three important movements concerned with central issues of contemporary society—women's role in society, environmental protectionism, and consumer advocacy—achieve their aims largely by depending on the unique character of the U.S. litigation system.

The early influential books of the modern feminist movement emphasized the urgent need for improved female contraception.

Simone de Beauvoir's *The Second Sex* stated explicitly, and Betty Friedan's *The Feminine Mystique* implicitly, that a liberated woman must be in control of her own fertility. Probably most women will agree that the Pill, more than any other single factor, contributed to that aim. But an informed and highly motivated group of women—primarily American and, by world standards, exceedingly affluent—while emphasizing their abhorrence at male domination, also strongly criticized the Pill and frequently did so claiming to speak for women all over the world. They were concerned when the first epidemiological studies documented some of the Pill's less obvious side effects. Women, who earlier had objected to being used as human guinea pigs, now asked why the Pill had not been tested more thoroughly.

An undercurrent of such feelings persists. In the 1992 edition of *The New Our Bodies, Ourselves,* produced by the Boston Women's Health Book Collective, one can find it remarked that "the Food and Drug Administration [FDA] approved the Pill for marketing in 1960 without adequate testing or study. . . . The Pill became a gigantic experiment: within two years about 1.2 million American women used it. . . ."

Such large-scale, postmarketing "experiments" are unavoidable, however, and occur with every vaccine and drug to which a person will be exposed for long periods of time. Only medicines used to treat acute conditions and used over short time intervals can be effectively screened for most side effects during the premarketing, clinical test phase. With regard to the question of why it took so long to lower the initial high dosages of the progestational and estrogenic ingredients, it must be remembered that abortion was completely illegal at that time; experimenting with lower dosages might well have led to higher failure rates for which no alternative could be offered to the women on whom the new dosages were tested.

RETRENCHMENT OF PHARMACEUTICAL INDUSTRY

In 1970, 13 major pharmaceutical companies (nine of them in the United States) conducted research and development (R&D) in contraception; by 1987, the number had dropped to four (only one of them in the United States). Today none of the active progestational and estrogenic ingredients of the Pill is manufactured in the United States.

The withdrawal of the large U.S. pharmaceutical companies from contraception R&D has had three major causes. The first was the stringent animal toxicology tests (seven years in beagles and 10 years in monkeys) demanded in 1969 by the FDA in response to concerns about the long-term effects of steroid contraceptives. These requirements were not modified until 20 years later as a result of overwhelming evidence presented by foreign regulatory agencies and the World Health Organization (WHO) about the futility of such special "overkill" mandates.

Second, the impact of the congressional Nelson hearings conducted between January and March of 1970 was exacerbated by the commentary of self-interest groups and further sensationalized by the press, thus giving the contraception field an extremely poor image. The fact that the pharmaceutical industry chose not to testify before Senator Gaylord Nelson's committee, and the subsequent disaster of the Dalkon Shield intrauterine device (IUD), only aggravated the hostility.

The third blow, and in the end the most devastating, has been the changes in the litigious character of our society, especially where drugs and medical practice are concerned. Unquestionably, the fear of litigation had a salutary impact on some practitioners and manufacturers in medicine in general and on birth control in particular. The Dalkon Shield is a prime example of a case in which litigation was essential. At the same time, contemporary tort law, with respect to legal liability, has altered medical practice for the worse.

With contraceptives, litigious practices have been extreme. In 1986, for instance, the Ortho Pharmaceutical Company lost a $5,151,030 judgment in Georgia because its spermicide Ortho-Gynol, used by a woman while she was unknowingly pregnant, was alleged to be the cause of her baby's birth defects—a possibility that is not consistent with current epidemiological evidence. And although in most malpractice and product liability cases (e.g., that of asbestos) the plaintiff recovers no more than one-third of the financial judgment, the remainder being consumed by the legal community, such litigation has added an enormous financial burden to precisely that segment of the population that the legal system was designed to protect: the consumer.

The impact of litigation on the Pill is especially instructive. Indisputably, some women have been physically harmed by the Pill, and it is reasonable for society to compensate them in one way or another. Even though few Pill suits that have gone to trial have been won by the plaintiffs, the legal defense cost for the drug and insurance companies has escalated to such an extent, especially because of liberalized discovery rules permitting plaintiffs' attorneys to demand tens of thousands of documents, that out-of-court settlement of such litigation is often cheaper than defending it in court. The Office of Technology Assessment (OTA) in its 1982 report stated that liability costs in the oral contraception field were higher than for any other drug category.

These legal costs are in the end paid for by the millions of women who benefit from the Pill and who would probably object greatly if they were returned to the narrow contraceptive options of pre-World War II days. Over the course of the past two decades the cost of a monthly regimen of the Pill has increased tenfold in the United States, even though most Pills now on the U.S. market have been "off patent" for many years. Fear of litigation and unavailability of insurance has eliminated market competition: until 1988, no generic versions of the Pill were available, and even the ones that have appeared recently cover only a fraction of the market and are, in any event, manufactured by the producers of the proprietary formulations.

PILL USE IN THE UNITED STATES

No new active ingredients have appeared in the Pills sold in the United States since the 1960s.[1] By contrast, three new ones (desogestrel, norgestimate, and gestodene) were introduced in Europe in the 1980s. The leading European manufacturer of the most advanced Pill has so far not introduced its product (desogestrel) in the United States—in part because of potential liability exposures. Yet this product has one of the lowest dosages of all Pills and, moreover, has an improved metabolic profile compared to the other progestational steroids currently available to women in the United States.

The negative publicity of the Nelson hearings resulted in both justified and unjustified caution about the Pill. Consumption dropped by over 20% to about 8 million women in the United States in the 1970s (although it continued to increase in Third World countries) but then started to rise again to the current all-time high of more than 13 million American consumers. The consensus now is that for healthy young women, the Pill is the most effective contraceptive method and probably one of the safest. Women in their middle-thirties or older were thought to be at increased risk in terms of cardiovascular complications, and the current pattern of use among such women in the United States reflects these beliefs, although the most recent epidemiological evidence concerning low-dose Pills suggests that such risk applies only to heavy smokers. As a consequence of these concerns, and because of the lack of other

[1]Some of the newer European formulations were finally brought to the U.S. market in 1993.

effective alternatives, the incidence of sterilization has risen so sharply (in contrast to Western Europe) that this essentially irreversible method now surpasses Pill use among married couples in the United States.

The attitude of feminist activists toward the Pill has also changed. Although one can still find occasional anachronisms like the born-again contraceptive fundamentalism expressed in 1984 by one of the early feminist writers, Germaine Greer (who indicates that she has no use for the Pill and even denigrates the diaphragm in favor of coitus interruptus, the cervical cap, and condoms), the current position of most informed feminist spokeswomen toward contraception in general, and the Pill in particular, reflects the realities of today. Like the vast majority of American women, they want for themselves and for their partners more choices, to suit the personal and professional lifestyles of women working outside the home. They want full and up-to-date information on each method. For the Pill, this includes dissemination of the potential negative side effects as well as of the more recently discovered noncontraceptive benefits.

Women are now represented in substantial numbers in decision-making bodies dealing with contraception, such as the advisory committees of the FDA, the National Academy of Sciences, the National Institutes of Health (NIH), and the World Health Organization (WHO). Also, whereas in the 1960s the overwhelming majority of American obstetricians and gynecologists were men, more than half of the residents and young practitioners in that subdiscipline are now women.

Yet just as women have entered every aspect of contraceptive development—from research and testing to delivery of the product—their choices are becoming more limited. This is primarily a consequence, again, of public, governmental, and media response to the complaints in the 1960s and 1970s of women who wanted a perfectly "safe" Pill or other contraceptive. What do the professionals in contraceptive research have to offer in that regard?

THE CURRENT CLIMATE

The fashionable area of human reproductive biology is now the study of infertility rather than contraception. The lessened prestige of the latter field is reflected by the paucity of new talent entering it. This situation exists partly because relatively less money is now dedicated to contraception R&D than in the past. Not only have industrial expenditures virtually ceased, but the principal U.S. government funding agencies, the NIH and the Agency for International Development, because of mandates initiated under the Reagan and Bush Administrations, were prevented from supporting many important areas of contraception research: To convert promising laboratory discoveries in animal reproduction into viable methods of human birth control is now so time-consuming, and so dependent on the participation of the pharmaceutical industry, that many scientists have turned to other fields because of the lack of material and societal support.

Another reason that scientific attention has turned away from contraception research is that, since the late 1960s, country after country in the developing world has recognized the problems of uncontrolled population growth and has started to implement birth control programs—some of them, such as the one in China, on a huge scale. Health professionals dealing with the delivery rather than the creation of contraceptive methods then decided that the emphasis in these countries should be placed on education, on the creation of the appropriate infrastructure, on the integration of contraception with maternal and child health care, and on the optimum use of existing methods (the Pill, IUD, condom, injectable steroids, and sterilization) rather than on the search for new contraceptive methods.

This focus of Third World governments suggests how different is the perspective of women in the United States compared with that of women in poor countries. Here, IUDs have been rejected by many women, largely because of the defective Dalkon Shield. Makers of other IUDs—the Ortho Pharmaceutical Company with the original Lippes loop and G. D. Searle with the Copper Seven—have also withdrawn them without any pressure from the FDA. IUDs never did play a role, in fact, in the single most important birth control issue in the United States, teenage pregnancy, because the device is unsuited to young, nulliparous women. Yet in China at least 35 million women are estimated to be wearing an outdated metal IUD developed in the 1960s, thus making it the most prevalent contraceptive in that country. In Mexico, similarly, where the government switched in 1974 from a laissez-faire pronatalist policy to an increasingly aggressive population control program, steroid contraceptives and IUDs are the key components of that program, followed by abortion.

In some Latin American countries, such as Brazil, IUDs are hardly used, and the Pill continues to be the method of choice, whereas many Asian women prefer steroid injectables, which certain women's health groups in the United States continue to oppose. All this proves that couples all over the world need more choices.

PROGNOSIS FOR NEW DEVELOPMENTS

The 1982 OTA report "Future Fertility Planning Technologies" introduced a list of future contraceptive methods by saying that "between now and the end of this century, more than 20 new or significantly improved technologies for contraception are expected to become available." A similar article, published by family planning professionals in 1986 under the title "The Next Contraceptive Revolution," gave virtually the same list and cited a lack of financial support as the chief obstacle to its immediate realization.

My own view is much more pessimistic; regardless of the amount of money available, none of the truly revolutionary developments such as antifertility vaccines or a male Pill have a chance of being used by the public in this century. The rest of the cited contraceptive improvements, which include another vaginal spermicidal tablet, another copper IUD, and a cervical cap, although clearly useful in a public health and demographic context, are not new or revolutionary. A delivery system for steroid contraceptives, which replaces the daily ingestion of a tablet by steroid-loaded vaginal rings or subdermal implants, is no consolation to women wishing to abandon continuous exposure to a potent steroid hormone, especially when they learn that these supposed novel developments have been underway for nearly 20 years.

A PRIORITY LIST OF NEW CONTRACEPTIVE METHODS

What new contraceptive methods are needed, and who would their principal beneficiaries be? The following list is short, yet ambitious.

1. *A new spermicide with antiviral properties.* The AIDS epidemic alone justifies putting this item on top of the list. Demonstrating antiviral activity, however, is not sufficient. A drug or formulation needs to be devised that will be effective under conditions of normal use during coitus. The noncontraceptive benefit is likely to weigh heavily in any risk assessment and in FDA approval.

2. *A once-a-month Pill effective as a menses-inducer.*

3. *A reliable ovulation predictor.* Such precise prediction is now technically feasible. What still remains to be done is to convert this into a financially realistic and operationally practical method for routine birth control. This approach to contraception would be equally attractive to prochoice and anti-abortionist groups.

4. *Easily reversible and reliable male sterilization.* At present, vasectomy can be reversed only through expensive microsurgery. Even when normal sperm count is restored, immunological reactions frequently lead to

infertile sperm. In the absence of virtually guaranteed restoration of fertility, presumably on the basis of epidemiological studies covering a minimum of two decades, the prospects for widespread dependence on vasectomy reversals are small, whereas the opportunities for malpractice litigation seem limitless.

5. *A male contraceptive pill.* In 1970 I documented the technical reasons why developing a male Pill would take longer than work on a new female Pill. Long-term assurance of safety, which has been insisted on by women for the female Pill, is only available through large, long-term epidemiological studies. Safety may be more difficult to establish for men than for women, primarily because of the longer fertile lifetime of men.

6. *Antifertility vaccine.* In principle, this would be the most revolutionary development; it would radically change our perception of human fertility if teenage males or females, or both, were vaccinated so that they would be infertile until a conscious step was taken to achieve fertility. To accomplish prompt restoration of fertility, a method would be needed that actively reversed the immunological infertility—before vaccination wore off with time. It will take many years of carefully controlled studies with large numbers of women volunteers to determine how long it takes for the effect of the antifertility vaccine to wear off, whether all women are then able to produce normal babies, and whether there are serious side effects after extensive use of such vaccines.

CURRENT BARRIERS TO CONTRACEPTION R&D

If only these six projects, and no others, were completed successfully, the choice for human fertility control would be vastly expanded for all constituencies—poor and affluent, prochoice and anti-abortion, female and male. What are the chances that this can be accomplished? The following analysis is presented primarily from a U.S. perspective, but it has global ramifications.

Two of the six agenda items—a new antiviral spermicide and a reliable ovulation predictor—require only the conventional incentives of the marketplace. There is no dearth of market incentives for antiviral drugs, and research in this field is burgeoning. If use efficacy against the AIDS virus can be demonstrated, the FDA is likely to expedite marketing of such an agent. An ovulation prediction test faces only straightforward FDA barriers, typical of any new diagnostic method, and no toxicity expenses (only a few drops of urine, saliva, or blood are required).

The other four approaches, however, have certain handicaps, the elimination of which will require major legislative or social changes. They will be used by healthy people—a circumstance for which society tolerates very little risk. Development times covering one or two decades make any investment extremely risky, be it a company's money or an investigator's career. If initiative and support were to depend on nonprofit or governmental agencies, long-term commitments would have to be made. Until now, only large, multinational pharmaceutical companies have had the resources and expertise for drug developments of the magnitude of these four contraceptive approaches. Given the long development time, recovery of the investment and generation of profit require a long proprietary position, which the present patent laws do not offer. The legal exposure to liability suits could be extremely risky. Impotence or prostatic cancer—two conditions commonly associated with aging in males—are likely to attract litigation by men who may take their Pill for 40 years and then blame it for their misfortune. Cases of permanent sterility would likely be attributed to an antifertility vaccine if millions of nulliparous women opted for such a method of birth control. Large pharmaceutical companies, selling many products and having many stockholders, are likely to be more sensitive to threatened boycotts and political pressures than smaller companies. A menses-inducer may well fall victim to a fear of such pressures, even though it may be the most efficient way to reduce abortions.

How can the hurdles be cleared that now stand in the way of a contraceptive revolution or even of modest progress? History demonstrates that no major advances will occur without the participation of the pharmaceutical industry—in production, distribution, development, and even research. The idea that such decisions should be left to the marketplace is useless, for the market has already spoken: Given the cost, time, and litigation risks, it is not worthwhile to invest in the development of new contraceptives. A survey of leading R&D therapeutic categories for 1988 does not list contraceptives even among the first 35 rankings. If society wants a well-stocked contraceptive supermarket, society will have to provide the impetus. Of all incentives, addressing the litigation problem in the United States would be of overriding importance. In fact, this is precisely the area where some moves have finally been initiated by legislators prompted by the crisis in vaccine production and the even bigger need for R&D of new vaccines for infectious diseases, from AIDS to malaria. Strangely, the similarity in the problems faced by the developers and producers of vaccines and most contraceptives has not yet been recognized in legislative circles.

MODIFICATION OF PRODUCT LIABILITY

The National Childhood Vaccine Injury Act of 1986 was introduced by Congressman Henry A. Waxman and constitutes a form of no-fault insurance against possible injuries from the seven pediatric vaccines. The rationale for this limitation was that all children must receive such vaccinations to attend school and that a few are bound to be harmed by such compulsory vaccination. Manufacturers of these vaccines threatened to withdraw from the field because of ever-increasing liability suits, and the Waxman bill was designed to stem such a crisis in vaccine production. Revenues come from a special tax imposed on any childhood vaccine. Like any insurance system, the beneficiaries are expected to pay the premium for risk protection.

The Waxman bill does not address itself to the problems of other vaccines, whose administration is not obligatory, nor does it pay attention to the even more serious problem associated with the development of new vaccines. It took another national calamity, the AIDS epidemic, and the public's interest in an AIDS vaccine, for the California legislature to examine remedies to the product liability barrier standing in the way of vaccine research. The bill introduced by Assemblyman John Vasconcellos is specifically limited to AIDS and applies only to California. However, it represents a promising model for federal legislation covering other vaccines and, as I wish to emphasize, also for contraceptives.

The key provisions of the California bill are that such a vaccine is recognized as "unavoidably unsafe" and thus exempt from strict liability lawsuits. The bill's key feature of restricting the manufacturer's liability and of funding compensation for medical costs, loss of earnings, and pain and suffering out of an extra charge imposed on the price of the vaccine would be essential in the contraceptive field. Improvements can be made in these legal models, notably a further restriction of tort law application, which is opposed by trial lawyers' lobbies, as are most other provisions of these bills.

Contraceptives and vaccines are obvious targets for superlitigation, because they are not curative drugs to be taken by people already ill; they are administered to healthy people to prevent a condition that the person may never get. Even though a no-fault insurance program, structured around self-funding, would be the single most important incentive for the gradual reentry of the pharmaceutical industry into the field of contraceptive innovation, there are differences in perception between vaccines and contraceptives that operate against extending any special incentives to the latter class.

The societal and personal costs of an undesired pregnancy and of an unwanted child are simply not equated by the public to the immediately evident health consequences of a disease, be it measles or AIDS. Among some groups in the United States, contraception is inherently suspect because of its actual or perceived effect on sexual mores. Finally, U.S. society is likely to look askance at incentives that, directly or indirectly, may benefit pharmaceutical companies, when such firms are generally among the most profitable sectors of U.S. industry. But when another decade or two of minor improvements of existing methods, or even of diminished contraceptive choices, has passed, and the number of abortions, legal or illegal, has not dropped significantly, when this bitter pill is tasted by the next generation, then the time may be ripe for substantive changes.

CONCLUSION

In view of the present political and social climate in the United States, and the minimal participation of the pharmaceutical industry in contraceptive development, all we can expect well into the beginning of the 21st century are minor modifications of existing methods: different delivery systems for steroids, possible improvements in sterilization techniques and barrier methods, more precise indications of the safe interval, and possibly a more realistic reconsideration of the IUD option. Such modest developments will extend contraceptive use patterns, but they will not affect our total dependence on conventional 19th- and 20th-century approaches to birth control.

Scientific Cooperation and the Developing World

12

Injectable Contraceptive Synthesis: An Example of International Cooperation

The original article, on which this revision is based, was coauthored with Pierre Crabbé and Egon Diczfalusy.

In the 1970s the limited number of preparations available for use as long-acting injectable contraceptives—a high priority item of family planning centers in many developing countries—and the lack of interest on the part of the international pharmaceutical industry in pursuing additional research in this area made it desirable to find a new avenue for the development of such substances. Comparative multicenter trials conducted by the World Health Organization's Task Force on Long-Acting Agents for Fertility Regulation in different countries revealed that, for existing formulations, the duration of action was only two or three months, whereas a duration of effect of up to six months was desirable. Therefore, the Special Programme of Research, Development, and Research Training in Human Reproduction of the World Health Organization (WHO) initiated a program for the chemical synthesis and screening of a large number of steroid derivatives with the aim of developing several long-acting formulations (monthly, three-monthly, and six-monthly).

At a meeting at the WHO headquarters in Geneva in January 1975 attended by Egon Diczfalusy (Karolinska Institute, Stockholm), Josef Fried (University of Chicago), and me, it was concluded that the development of a new, long-acting contraceptive agent would be worth combining with an effort at institution building in lesser developed coun-

Note: Original text written in 1980.

tries (LDCs). In addition, it was believed that this program might serve as a model for other drug development programs outside the traditional pharmaceutical industry. For example, the development of drugs for certain parasitic diseases had been neglected by the pharmaceutical industry; a similar approach could also be envisaged for the creation of new pesticides.

In July 1975 eight internationally recognized steroid chemists and endocrinologists with past or current experience in the pharmaceutical industry attended a meeting held under WHO auspices in my office at Stanford University. The chemists of the group compiled a list of approximately 150 hypothetical steroid compounds that they considered could be synthesized and should be subjected to biological screening in a program designed to uncover new and effective sustained-release injectable contraceptives. They also proposed 15 laboratories as candidates for participating in the program to synthesize new steroids. These laboratories, most of which were in developing countries, were contacted by WHO headquarters staff to determine whether they would be receptive to the idea of participating in the program.

The arrangement proposed was that WHO, in addition to supplying literature, material, and chemicals, would fund each laboratory with $10,000 to $15,000, the sole requirement being that 5-gram quantities of pure steroid would have to be delivered to WHO headquarters. Patent rights would remain with WHO.

CHEMICAL OBJECTIVES

The objective of the chemical synthesis program was to develop new fertility-regulating agents in the following areas: 3–6 months progestogen-only injections, monthly estrogen–progestogen injections, and long-acting testosterone–progestogen combinations. The common denominator in these agents was that they would be administered by injection and would be derivatives of synthetic steroids such as norethindrone and levonorgestrel, which were known to be efficient and "safe" as contraceptive agents.

It was considered at the outset that this objective might be achieved by chemically modifying an active contraceptive steroid drug into a "prodrug" that would either be inactive or less active than the parent steroid. When administered to humans, such a prodrug would be converted into an active contraceptive agent by enzymatic hydrolysis in vivo. The rate at which the hydrolysis occurs determines whether the prodrug might be suitable for use as a long-acting, injectable contraceptive.

The main goal of the program initiated in 1976 was to design novel steroid esters that could serve to enlarge the number of long-acting injectable contraceptives available to women. The strategy was typical of that used frequently in industrial organizations in that it involved the initial preparation of a number of esters of the known contraceptive agents 17α-hydroxyprogesterone, norethindrone (norethynyltestosterone), and levonorgestrel.

The natural male sex hormone, testosterone, was also included in the program to provide eventual candidates for an injectable male contraceptive. Norethindrone was selected as a potential progestogen because laboratory and clinical experience had shown that it was one of the safest progestogens available and was no longer protected by patents. Levonorgestrel, although still covered by patents, was chosen because of its high progestational potency.

ORGANIZATIONAL ASPECTS

Eventually, 12 laboratories agreed to participate in the program. The laboratories were located in Australia, Brazil, Bulgaria, the former German Democratic Republic, Iran, Mexico, Nigeria, Poland, Singapore, Spain, and Sri Lanka. Participants each received a list of 16–20 ester structures and were invited to submit research proposals outlining how they intended to synthesize the acid chain to be introduced into the steroid molecule. They were also asked to submit budgets. These items were reviewed by the WHO Secretariat with the aid of at least one outside referee and the coordinator[1] of the program. Once the proposals were approved, a Contractual Technical Services agreement was drafted, which permitted WHO funding to be instituted.

During the first three years of the program the coordinator visited most laboratories at least once. The purpose of the site visits was to brief the investigators about the precise objectives of the program and operating details, to offer appropriate scientific information and advice, to become familiar with the local research facilities, and to solve a number of administrative problems. Each center had different difficulties: for example, inexperienced personnel, lack of sophisticated equipment and instruments, inclement weather conditions, or frequent power failures. Most centers faced one common problem: complying with the

[1]Throughout the entire program, the coordinator function was exercised by a former Belgian postdoctorate fellow of mine, the late Pierre Crabbé, who followed me to Syntex and for a time served as director of chemical research for Syntex's Mexican research division.

cumbersome and time-consuming regulations that many of the countries imposed on importers of chemical reagents and small equipment items.

It soon became evident that many of the developing countries would do well to overhaul their customs regulations if they wished to upgrade and expedite scientific research. The only feasible solution for the chemical synthesis program proved to be for WHO to keep in Geneva part of the money allocated to each center. The WHO Secretariat ordered directly the chemicals for the principal investigators in accordance with their requests and budgets, and arrangements were made to ship the chemicals through the auspices of WHO international channels. The organization of such a network was difficult, but in the long range it proved to be crucial because it saved a great deal of time, unnecessary delays, and frustrations. (This modus operandi even made it possible to maintain a productive steroid synthesis program under A. Shaflee at the University of Teheran during the entire Iran–Iraq war.)

In addition to providing background information and technical literature to all principal investigators, the coordinator was in charge of examining and labeling the samples (a code number was used for every compound submitted by the centers) and checking the data sheets (each sample sent to WHO headquarters was accompanied by a data sheet giving a full description of the substance, including its physical, spectroscopic, and analytical properties). The coordinator acted as a general troubleshooter and was easily available to the various principal investigators; together with the WHO Secretariat he reviewed all the manuscripts originating in the various centers because of their possible impact on the filing of patents on behalf of WHO for the most promising compounds.

The preparation of the long-acting esters presented several difficulties. First, the acid chains—encompassing more than 100 different chemical structures—had to be synthesized, because most of them had not been described in the chemical literature. Their synthesis invariably involved a number of steps, frequently with a variety of stereochemical complications. Another problem was the esterification reaction, which is difficult to perform on tertiary alcohols of the type present in norethindrone and levonorgestrel. In several instances, sophisticated techniques were used and with certain acids a new esterification method had to be developed. A third problem, both time- and material-consuming, was that of reaching the requested high purity (about 99%).

FORMULATION AND BIOASSAY

After they were synthesized, all compounds were shipped to the Department of Chemistry of the City University, London, for quality control and further purification, when necessary. The compounds were then forwarded to the School of Pharmacy of the University of London for formulation. Once satisfactory stability was established, the microcrystalline suspensions were sent for bioassay to the National Institute of Child Health and Human Development of the U.S. Department of Health, Education, and Welfare in Bethesda, Maryland. This unit of the National Institutes of Health (NIH) handled both the funding and the operational details of the biological evaluation in rodents. Esters not suitable for the preparation of aqueous suspensions were bioassayed as oily solutions. After leaving WHO headquarters for quality control at the City University, London, every sample was followed by the WHO Secretariat and by the coordinator (appropriate forms were used at every stage of the program) until the biological results were returned.

RESULTS

The synthesis and screening programs were reviewed in depth during consultations held between November 1977 and January 1980 with the participation of most principal investigators, expert chemists, biologists, the coordinator, and WHO staff members. Approximately 220 steroids were synthesized in the 12 participating laboratories, many of which had had no previous steroid experience.

The biological evaluation of potential male contraceptives was discontinued for financial reasons, but six compounds of possible use in female contraception were selected for further development. The next phase of research—testing in primates and preparation of larger amounts for toxicology and for eventual phase I clinical studies in one of the WHO Clinical Centers—was then initiated.

Excluding efforts by military establishments, this was probably the first instance in which an international public sector agency launched successfully a program of this nature, and it is reasonable to ask how economical the program was. In terms of time, the chemical synthesis took considerably longer than it would have in the steroid synthesis laboratory of a large pharmaceutical firm in the United States, Western Europe, or Japan. However, part of that extra time was consumed in institution building and in creating technical capability in developing countries—two features that were of long-lasting benefit. In terms of direct funding from WHO (and indirectly from NIH through its support of the biological screening), this program proved to be much cheaper than would have been the case in a pharmaceutical industrial laboratory. All of the indirect costs and a substantial portion of the personnel charges were absorbed by the participating university and government laboratories, thus making this a truly cooperative economic project. However, if these indirect costs had been combined, then it is unlikely that the program would have been much cheaper than the usual industrial effort. What is important is that societal goals rather than pure economics were the driving force.

CONCLUSION

Even if no practical new contraceptive agent is developed as a result of this research effort, the WHO program illustrates how a multinational cooperative project in drug chemical synthesis can be organized outside the traditional pharmaceutical channels—a model of particular relevance to lesser developed nations. The program is an impressive example of an interdisciplinary and fruitful cooperation between an international organization (WHO), a prominent national public institution (NIH), and university or government laboratories from countries all over the world.

13

Future Methods of Fertility Regulation in Developing Countries: How to Make the Impossible Possible by December 31, 1999

For once, the topic of major innovations in fertility control should be put into the proper perspective by focusing on the priorities of the lesser developed countries (LDCs) rather than continuing to look at it through the inverted telescope of occupants of the wealthiest countries (MDCs, or more developed countries). The present modus operandi in MDCs is unlikely to lead to significant breakthroughs in *practical* fertility control by the end of this century.

SPECIAL REQUIREMENTS OF LDCs

In the MDCs it is readily conceded that the quality and variety of birth control deserves substantial improvement (through the introduction of several novel approaches to contraception) and that the needs of certain segments of the population are poorly served (e.g., the teenage pregnancy rate in the United States and the extremely high abortion incidence in Eastern European socialist countries). Nevertheless, no political or economic disaster would occur if not a single improvement

Note: Original text written in 1983.

in currently available fertility control were introduced in those countries by the end of this century. In other words, the relative low population growth rates in those MDCs would continue.

By comparison, let me list just a few problems peculiar to LDCs. Although not all of them apply to all LDCs, primarily because "LDC" or "developing country" are relative terms, many of these problems are prevalent.

1. The population growth rate is much higher (than in MDCs), and in some countries it is still increasing. Consequently, the political and economic consequences of population growth are much more severe in LDCs than in MDCs, especially if one considers the enormous differences in economic resources (e.g., per capita food production in sub-Saharan Africa has dropped consistently since the 1960s; yet within the next quarter of a century, its population will more than double).

2. Of the relatively few birth control methods employed in the world, even fewer are applicable (partly for operational and partly for cultural reasons) to large segments of a typical LDC population (the People's Republic of China would appear to be the most striking exception to this generalization).

3. Few LDCs, and for that matter few MDCs, have established a wish list of what type of future birth control approach would be particularly suitable for a given population, taking into consideration cultural, logistic, and economic realities of the country.

4. Even if such a wish list existed, most of the LDCs could not implement it for want of scientific, technical, and financial resources.

5. Many of the LDCs have no or only a rudimentary regulatory agency (of the U.S. Food and Drug Administration [FDA] type) and thus are dependent, implicitly or explicitly, upon the actions of the regulatory agencies of MDCs (such as the United States), which are not concerned with the question of what birth control procedures are needed elsewhere and what risk–benefit determination is applicable in such countries. One common denominator of government regulatory agencies in technically sophisticated and economically advanced countries is their extreme parochial view, which is particularly pervasive in birth control because supposedly one is dealing with "drugs for healthy people." (Is a pregnant 14-year-old girl or a 26-year-old impoverished woman with five children "healthy"? Should the quality of life of such women not also be part of the definition of health?) Even Donald Kennedy, probably the academically most sophisticated and societally most compassionate of FDA commissioners, felt it necessary to state in congressional testimony in 1978, "The

question is this: Should national drug approval decisions in the United States be influenced by the needs of users of those drugs abroad? And if you answer that question 'yes', I submit that . . . you would be on the slipperiest public policy slope you can possibly conceive of. You would have to make a global cost–benefit decision for every drug which you are then approving for use in a single national population. I know of no way, nor do my colleagues know of any way, in which such a process could be conducted rationally."

6. Maternal death rates associated with childbirth in the poorer LDCs can be astronomical (i.e., 290 per 100,000 women in the 15–19 year age group) compared with the average in MDCs (i.e., 5.6 per 100,000 women of the same age group).

7. Time is particularly expensive for LDCs (*see* the following section).

REGULATORY CLIMATE AND ASSOCIATED COSTS

When dealing with research and development (R&D) of new human contraceptives, the time component is fantastically expensive. Money, on the other hand, is almost ludicrously cheap. For instance, in 1979, the worldwide annual expenditures for reproductive research amounted to approximately $150 million, which corresponded to approximately *two hours* of world armament expenditures.

Expenses can be calculated in many different ways, whereas this is not true of time. In government and academic institutions, the true indirect costs of R&D are usually underestimated or ignored. In industry, the converse occurs frequently. However, the biggest uncertainty in determining costs lies in the indirect *scientific* resources of the institution conducting the R&D operation. Starting a new entity is much more expensive because one needs a large critical mass of multidisciplinary scientific personnel and associated laboratory equipment, and most importantly, *accumulated experience* in the particular fields.

The intelligent government or funding agency should decide where the necessary work can, in fact, be done in the cheapest manner. At times, factors other than costs are important for making such a decision, a typical example being the 1975 chemical synthesis program sponsored by the World Health Organization (and supported by the National Institutes of Health) for the development of injectable contraceptives. This multiannual program took into consideration the reluctance of the international pharmaceutical industry to pursue research on improved injectable steroids—a high priority item of many LDC health officials—and the importance of institution-building in LDCs.

The biggest unsolved problem, and the most time-consuming one, is the attempt to minimize and predict risk in human subjects through appropriate animal models. A typically rigid approach was exemplified by the FDA's animal toxicity requirements initiated in 1969 for conducting clinical work with contraceptive steroids. These requirements, which specify the animal species to be used, were not changed for 20 years, in spite of the hue and cry among professionals and even many regulatory agencies in other countries about the obligatory use of dogs (beagles), probably because nobody had come up with a face-saving formula whereby a government agency can gracefully admit to having made a judgmental error. (New guidelines initiated in the 1980s for other approaches appear to be more flexible in the sense of leaving more choice for the experimental animals and reducing the time.)

REDUCING TIME AND COST REQUIREMENTS

In my opinion, during the next two decades, distribution and utilization of *existing* birth control approaches will be improved considerably. However, it is unlikely that by the end of this century dramatically new approaches to human fertility control will be available to the public in LDCs unless we take heroic steps to reduce the time interval needed for the development of such approaches. I am convinced that if the time can be reduced, money will become available—most likely even from industrial circles—to pay for such R&D. Five proposals should be consistent with the following premise, which applies to biomedical research in general but to research in fertility control in particular: Let us expedite research on societally beneficial and desirable innovations with maximum feasible protection of the consumer. Are these two criteria, time saving and consumer protection, mutually incompatible, or can a regulatory climate be created that is supportive of such an aim? As a minimum, regulation means delay. Some of these delays are indispensable and unavoidable; others are unjustified. At its ultimate, some regulation, notably in the field of fertility control, means not even starting some research that eventually may lead to very important societally beneficial agents.

1. *Expedite clinical research phase.* Given the fact that in reproductive biology animal models are particularly difficult to find in order to predict human clinical results, it is essential that many new approaches be brought as quickly as possible to clinical experimentation. Given the relative paucity of leads, of different approaches, and (in the case of synthetic agents) of fundamentally different chemical entities, what is

needed most for rapid progress in the field of human fertility control is clinical evidence that certain novel approaches have a sporting chance of success.

I recommend a flexible approach (*see* recommendation 2) that will depend in part on the time that a clinical subject will be exposed to the drug, but that generally will only require subacute toxicology in one to two species for maximally 30–60 days. During this clinical research phase one will have to depend largely on Institutional Review Boards (IRBs) and on Good Laboratory Practice (GLP) in those central facilities that will be used for laboratory work. Where IRBs do not exist, it will be necessary to establish them. It will be further necessary to work out a good spot-checking and monitoring mechanism for IRBs and GLP, but otherwise the modified requirement should make it possible to initiate clinical work as rapidly as possible.

In the past, during most of the significant research and development on improved contraception, remarkably little irreversible harm had been done to clinical subjects during the initial investigational phase. Whatever serious problems have arisen have occurred almost invariably at a later clinical stage when the product was used widely with much less or no monitoring.

2. *Extreme need for flexible and ad hoc guidelines.* Understandably, the developer of new methods, especially if it is a pharmaceutical company, would like to have precise guidelines in advance so as to be able to estimate the cost of development and to be sure that some time-consuming requirements are not brought up at the last moment. Equally understandably, the promulgator of these guidelines tries to anticipate all conceivable eventualities that need to be taken into consideration. The consequences are rigid and often cumbersome rules. Because the crying need in the fertility regulatory area is greater diversity and novelty, many requirements cannot be anticipated. Although some new ones may have to be introduced, others may be totally redundant. An international body of experts, rather than bureaucrats, should be created that meets frequently and that will establish guidelines for each new (in the truest sense of the word) proposal using both scientific facts and especially the priority system listed in section 4.

3. *Protection of human subjects.* A mechanism has to be devised that is both humane and efficient as far as the protection of the human subjects is concerned. Here I am speaking both of subjects during the clinical experimental phases and of the final consumer. Protection of the former

is simpler because they are fewer in number, they are better monitored, and they can be better informed. A reasonable government-subsidized insurance system should be instituted for persons involved as clinical experimental subjects. A number of proposals can be made to protect the ultimate consumer of such novel birth control procedures without resorting to the extremely litigious system currently operating in the United States and likely to spread to other MDCs, which seems to be based primarily on motives of greed and revenge rather than on help for the victim and/or prevention against re-occurrence of the particular event.

4. *Priority rating.* This is the most difficult and also most radical component of my proposals. Time and cost should really be a quantitative reflection of a society's risk–benefit evaluation. Virtually all decisions pertaining to time and cost are made by the MDCs and, therefore, are based deliberately or unconsciously on the priority of the MDC. A system should be worked out whereby the priorities of the LDCs can be expressed at least semi-quantitatively. Such a priority rating should imply that every effort would be made to reduce the time (i.e., willingness to tolerate initially greater risks) from laboratory discovery to ultimate introduction to the public, whereas the actual monetary price to be paid would be relatively unimportant because the ultimate benefits—notably the economic ones to society—would be so overriding. Components for the priority score should include such items as fundamental novelty of approach selected, degree of acceptability for widest range of given population, extent of new and expensive infrastructure needed to promote or utilize a new approach, likelihood of technical success, minimal irreducible time needed to successfully complete the project (this factor alone may place many promising approaches near the bottom of the priority scale), and inherent risks of method.

5. *International approval mechanism for fertility control research.* It is unrealistic to assume that any of the regulatory agencies of the highly developed countries, notably the FDA, would respond very positively to, or implement rapidly, the proposals made. The social climate in the United States and several other Western countries is currently not risk-oriented in biomedical research. Furthermore, birth control has a relatively low national priority in these countries because even the absence of any progress in the field during the rest of this century will have few serious effects in such MDCs. Finally, the more powerful a country is, the less likely is it to yield in any one area of national sovereignty.

Clearly, what is needed is an international technical body, and perhaps the World Health Organization (WHO) could serve this purpose. By limiting its responsibility solely to technical and clinical aspects of birth control, its mandate and mode of operation should be easier defined and implemented. Countries would join voluntarily and in so doing would agree that the approval function for the *clinical experimental* phase would be relegated to that international body (ultimate approval for commercialization of such a fertility control agent in a given country can always be left to the specific country). As more and more countries adhere to such a scheme; as experience is gained, and the world discovers that no avoidable disasters have occurred; as research developments under the new agency's responsibility are completed within a shorter time frame than currently occurring in MDCs, then it is conceivable that even MDCs may eventually join this system for approving clinical research in human fertility control.

CONCLUSION

It is easy to present many arguments why these recommendations will not work or are too utopian. It is more difficult to come up with constructive alternatives that will be responsive to the biggest problem that governments and planners have consistently ignored, namely the time factor. As a starting point I would propose a conference[1] that addresses itself to recommendation No. 4 (priority rating) in an unconventional way.

I would select knowledgeable representatives from at least 8–10 different LDCs in order to cover the range of economic, political, cultural, and religious differences that exist within the term "LDC." I would ask them to ignore, for the time being, technical realities and instead to present a wish list of what approaches to human fertility control would be particularly applicable and acceptable to their country. I would convert this into a global list to see which approach is acceptable to the largest number of people in *those countries that would be prepared to promote the use of such novel methods if they became available.* I would then ask another group of experts, this time selected purely on the basis of technical/scientific knowledge, rather than geographical or national background, to evaluate the feasibility of developing the two or three approaches with the highest priority score. I would ask an expanded

[1]A variant of this proposal was raised by me for consideration by the Pugwash movement (*see* Chapter 15).

group of such experts—now liberally supplemented by clinicians and public health professionals—to indicate the fastest way of bringing such a program to phase III clinical work, and I would then address myself, still in a hypothetical way, to designing clinical and regulatory protocols that would save the maximum amount of time commensurate with protecting in a realistic manner the subjects of such clinical research. Finally, I would compare the conclusions with the present state of affairs and if, as I anticipate, there should be a great difference between the two, I would encourage the creation of an international approval mechanism for clinical work on human fertility control agents along the lines indicated in proposal 5.

14

A High Priority? Research Centers in Developing Nations

From the standpoint of scientific development, a "developing" country becomes a "developed" one when original research emanates from it. The eventual consequence of such research is the creation of technological innovations, which may then be utilized in many other countries that have the manpower to accept such innovations, but not the technical ability to create them.

The ability to perform advanced research, be it in a university or other research center, usually appears at the end of the shopping list in educational and technological priorities of developing countries. The priority items are the rapid widening of the elementary educational base (frequently involved with an eradication of illiteracy), followed by the creation of universities and technological institutes whose main purpose it is to create technicians. This pool of technicians generates the teachers, civil servants, public health personnel, and applied technologists who are indispensable to the operation of any country. The creation of advanced centers for the training of scientists at the doctoral level with principal emphasis on basic research is usually considered a luxury that comes last, because of the pressure of the higher priority items mentioned. The few research scientists from such a developing country receive their advanced training abroad, and upon returning to their home base either find no opportunity to utilize their training or

Note: Original text written in 1968 and 1975.

else encounter conditions that are so inferior to the ones to which they became accustomed during their foreign training that they are not prepared to accept them. The result is the well-known "brain-drain"—either by emigration or by remaining in the home country but pursuing quite a different occupation. As far as research potential is concerned, this is equivalent to brain-drain by emigration, because with few exceptions a research scientist in an active field cannot be put on ice for a few years and then be expected to operate creatively after such a hiatus.

The so-called high-priority items of developing countries—the improvement in indigenous primary, secondary, and lower university education—have one thing in common; they take many years before noticeable improvements can be achieved. It follows then that even if the financial resources were available, the training of a pool of research scientists able to perform fundamental research would occur some time in the very distant future. If we now consider the ever-increasing tempo of scientific and technological development in the advanced developed countries, then it becomes obvious that the creation of basic research centers in a developing country through the traditional routes and time sequences is virtually a hopeless proposition. This is particularly true in scientific research where there exists only one standard of excellence, namely an international one. A hypothetical statement such as "this is very good chemical research for Kenya, but rather poor for Sweden" is equivalent to saying: poor chemical research is being performed in Kenya.

Without denying the crucial importance of raising the primary and secondary educational standards of a country, I feel that it is indispensable that a parallel effort be made to establish some centers for basic research of internationally recognized excellence for the following reasons:

1. By demonstrating in a few developing countries that internationally recognized basic research centers can be created without waiting for the logical educational and technological development of the country, an example is created, which (with possible modifications) may prove useful in other developing areas of the world.

2. By selecting a research field (typical examples will be given later) that offers ample opportunity for fundamental research and yet may have eventual practical technological consequences, the possibility of generating new industrial developments is presented.

3. By selecting a research field with a maximum eventual multiplication factor (e.g., chemical research, which requires microbiological, entomological, and pharmacological evaluations, and leads eventually to

clinical and veterinary applications), a fairly rapid and yet logical broadening in the scope of the research effort can be effected.

4. In countries where scientific research offers no status symbol whatsoever, a beginning will be made in this direction so that eventually scientific research may become a desirable career in that society.

5. An opportunity will be provided in the particular country for advanced training in certain research fields without the necessity of initially sending promising young students abroad. Once they have been exposed to serious research in their own country, subsequent foreign training (even in another field) becomes more meaningful and a return to the home country much more likely, because they will realize that research of an internationally accepted standard can be performed in their home country.

6. One factor in the complicated brain-drain problem—the unavailability of research facilities—will be partially eliminated in certain fields.

7. To raise the overall scientific level of a country or even of one university is difficult and time-consuming. To create selected centers of excellence in a few fields is easier, and their existence, in turn, frequently creates a stimulus in many other areas. Furthermore, the local "image factor," though of limited value, should not be ignored, because it is easier to point to a concrete and operating entity (e.g., the International Rice Institute in the Philippines) than to a statistically significant reduction in the literacy rate of a country.

The problems involved in establishing such a research center are numerous, and several will be peculiar to the particular country selected. Nevertheless, some general proposals can be put forward. Before doing so, I would like to offer a concrete example from my own personal experience that has some bearing on the hypothetical case that I shall present later.

CASE HISTORY: MEXICO

In 1949, when I, an organic research chemist, moved to Mexico for a limited period of time, several universities existed in the country, notably the huge and venerable National University of Mexico. It had a faculty of chemistry, which graduated several hundred chemists (Bachelor of Science level) each year. By American standards, the training and facilities ranged from mediocre to poor, but apparently they sufficed for the type of positions open to most graduates—analytical chemistry in industry, positions in small pharmaceutical laboratories, the nationalized petroleum industry, and so on. No graduate training in the American sense of the word was available, and no Ph.D. had as yet been offered in chemistry from a Mexican institution. Practically no basic chemical research was being performed in the country, and no fundamental chemical discoveries with a technological fallout had been made in Mexico. On the other hand, a sufficient number of technicians were available so that various foreign technological developments could be adapted locally, the manufacture of certain plastics being a typical example.

In 1949 the therapeutic effects of cortisone were discovered in the United States and received worldwide publicity. As a result, academic as well as industrial research teams in Europe and the United States started to work feverishly on developing new synthetic methods for this important hormone. In the same year, the hormone sales of Syntex (a small Mexican chemical firm, producing steroid hormones from yam-derived starting materials) had risen sufficiently that it could afford a research department of its own, and it was for this reason that I joined the company as associate director of chemical research.

Because steroid chemistry per se was not taught in Mexican universities, most of the Mexican technicians were trained on the premises, and a number of fellowships were established, whereby Mexican university graduates could do their research in steroid chemistry in the Syntex Research Laboratories. As a matter of enlightened self-interest, the company subsidized the operation of the Institute of Chemistry of the National University, so that in the early 1950s, the major portion of the operating budget of that Institute was based on annual donations from Syntex. As much of the advanced research had to be performed by Ph.D.-level chemists, who essentially did not exist in Mexico, most of the research chemists were brought from abroad, and by 1959, Ph.D. chemists from more than a dozen different countries were working at Syntex.

Scientific publication of research results in recognized international journals of the United States and Europe was encouraged—in fact so much so that by 1959, more scientific publications in steroid chemistry had emanated from Syntex than from any other academic or industrial organization in the world. Some of the research results also gained general publicity, for example, the first synthesis of cortisone from plant raw materials or the development of oral contraceptives, but the important fact is that in a matter of ten years, Mexico—a country in which no basic chemical research had been performed previously—had become one of the world centers in one specialized branch of chemistry. Other companies were also formed, and by 1960 well over 50% of the world's supply of steroid hormones originated from that country.

Other indirect benefits resulted. Although virtually no patents on steroids had previously been filed in Mexico, starting in the early 1950s a veritable flood of applications from all over the world were filed in the Mexican Patent Office. The Institute of Chemistry of the National University of Mexico, initially supported largely by Syntex and staffed almost exclusively by "graduates" of Syntex, received larger subsidies from government and philanthropic organizations and developed a Ph.D. program of its own. Many of the first doctoral theses dealt with various aspects of steroid chemistry and gained a well-deserved reputation in that field for the university.

The multiplication factor mentioned earlier is further demonstrated by the Mexican example in steroid chemistry. The chemical developments required extensive clinical screening, and a great deal of clinical endocrinology and steroid biochemistry was stimulated in the country as a result of the presence of a steroid chemical research center in Mexico. The industrial raw material, diosgenin, was derived from a plant collected wild in the jungles of southern Mexico. As a result of the ever-increasing demand for this raw material, extensive botanical surveys of that portion of the country were undertaken, and several companies, including one from Germany, established botanical research stations in an effort to develop a new cultivatable crop. The utilization of steroid hormones in animals (better food utilization in cattle, estrus synchronization in sheep, etc.) prompted a great deal of veterinary research all over the world, some of it also performed in Mexico.

Why did these developments happen in Mexico? It was a matter of proper timing, the existence of a domestic raw material, and some entrepreneurship. However, the plant material is also abundant in Central and South America, in South Africa, in India, and in China, and if the proper incentive and entrepreneurs had existed at that time in those countries, the same development could have occurred there instead.

The principal reason that I am citing this Mexican example in such detail is that it illustrates the key component of my proposal for the feasibility of establishing basic research centers in developing countries, namely an answer to the question, Who will do this research in the absence of the requisite trained manpower?

The local supply of Ph.D. chemists in Mexico was negligible, and virtually all of them had to be imported. Most of them were in their early twenties and had been trained in the major universities of Europe and the United States. Why were they prepared to move to Mexico? The salaries were just barely competitive with those paid in comparable American laboratories, but certainly not extraordinary. The physical laboratory conditions were good, but again not superior to anything found in the United States or Europe. In order to overcome the obvious isolation inherent in working in locations distant from other active centers of research, opportunities were offered to attend scientific meetings abroad. A few leading university professors from the United States, England, and Israel served as consultants and visited the Mexican laboratory every three months.

The excitement of doing research in a foreign country is of particular appeal to younger scientists, and this is precisely the group that we wished to attract. Even the scientific publicity value—a scientific publication of good quality from a scientifically unknown country is noted by one's peers much more readily than one emanating from Harvard or Oxford—was a positive factor in the considerations of these young scientists to move to Mexico. Finally, they were given the opportunity of having several local technicians helping them in the laboratory and, although this initially required personal training (with its own rewards), it enabled them to work with more "pairs of hands" than would have been possible for them in their home countries.

ESTABLISHING A CENTER OF EXCELLENCE

Let us now apply some of these conclusions to the problem at hand—the establishment of an internationally recognized center of excellence in some area of basic scientific research in a country that does not possess the trained manpower for performing such research. Some of the problems and possible solutions for them are the following.

1. Location: The research center must be close to an international airport. This requirement is not only for the convenience of the part-time directors (*see* Number 3 of this list) but is particularly important for the rapid delivery of equipment, spare parts, chemicals, and other materials on a continuing basis. There is no doubt that virtually all such items would have to be imported by air. The research center might not be part of a university, at least not during the first few years of its existence (*see* Number 4 of this list), but it should be located close to one, because of the great impact that it may have on graduate training in that university. The sharing of library facilities should also be of advantage. Suitable housing for the foreign staff (*see* Number 3), a reasonable

climate, and some element of political stability are other factors to be considered. A typical, but by no means unique example satisfying these criteria is Kenya, whose capital, Nairobi, is served by more than a dozen international airlines.

2. Selection of Research Field: This subject and one of staffing (*see* Number 3) are the most crucial. Initially, an international ad hoc committee of scientists from different disciplines should present a list of possible research areas, which should meet most of the following conditions: The field should be an active one in which there is a substantial international reservoir of trained young scientists at the doctorate level in order to satisfy the postdoctorate staffing mentioned under Number 3. Unusually high capital expenditures should not be required, thus excluding fields such as high-energy physics. Eventual long-term technological developments might emanate from such research (novel drugs or insecticides, novel food supplements, or new advances in electronics). Possible local advantages (availability of plant raw material for steroids in Mexico) might exist for conducting such research in a particular country.

With such criteria in mind, it would not be too difficult for such a committee to present a few reasonable conclusions. For instance, X-ray crystallography, a very active field of research, is dependent on large computers. The unlikeliness of technological fallout would presumably eliminate such a field. On the other hand, many aspects of oceanography, starting with organic chemical work on novel natural products from marine sources and ranging all the way to fundamental marine biological studies, could be a suitable field of research in the proper geographical environment.

3. Staffing: The basic assumption will be made that for all practical purposes no trained research scientists (as distinguished from technicians) are available in that research field in the country under consideration. The key question then is how to find the necessary scientific staff. I would like to propose an approach that has the virtue of using in a directorial capacity some of the leading experts in a given field, who under ordinary circumstances could not possibly be hired for such a position. At least in my own field of organic chemistry, there exists considerable precedent to indicate that the following approach would be feasible.

Let us assume that the specific research area has been selected and that 15 scientists at the Ph.D. level are required for the initial program. I suggest that they be divided into at least three groups, and that the research program of each be directed by a part-time director who will spend no more than seven days, three (or possibly four) times a year at the research center. The part-time director will be concerned with the overall research program of the group but will leave the day-to-day supervisory problems to a younger deputy. The five Ph.D.-level scientists working in the group should be postdoctoral fellows, who commit themselves to two-year (perhaps renewable for a third year) appointments. In areas such as chemistry, such postdoctoral appointments are the rule rather than the exception, and the manpower pool is an international one.

One may ask why such research fellows would wish to work in a research center in a "developing" country rather than in one of the major university centers in the United States or Europe. Judging from my own experience with a small group in Brazil, many young postdoctoral fellows are tempted to spend a couple of years in an exciting and, for them, different area of the world, provided it is in association with a well-known scientist, in order to be valuable for their eventual career placement. I can cite two very distinguished chemists in my own field who directed such research groups by long distance.

Professor R. B. Woodward, a Nobel laureate from Harvard, had a small postdoctorate research group in Basel, Switzerland, and Professor D. H. R. Barton (a Nobelist from the Imperial College of Science and Technology in London), one of Britain's most distinguished chemists, had a similar group in a private research institute in Cambridge, Massachusetts. Each of them crossed the Atlantic three or four times a year for short but intense personal visits, and some of the research results published by the groups were nothing short of spectacular. Needless to say, Woodward and Barton had no problems in finding suitable postdoctorate fellows from all over the world (I personally know of American, Swiss, English, German, and Indian members of their groups) for these small research groups, even though they were working in privately supported, non-university centers.

I venture to predict that no difficulties in such staffing would be encountered if individuals of such caliber were involved in part-time direction of postdoctorate research groups in a center located for instance in Nairobi, and similarly that no dearth of part-time directors would be experienced. (In connection with the problem of persuading potential postdoctorate fellowship candidates to spend one to two years in such a location under such circumstances, I posed this question to about 50 chemistry graduate students at Stanford University, using East

Africa as a typical location. Well over 50% of them indicated that they would be interested in such a position.)

The advantages of such an approach are many-fold. No single foreign country would have a predominance among the part-time directors or the postdoctorate staff, and turnover in the latter group would ensure no permanent foreign vested interests. On the other hand, within a few years, one would have created an international cadre of scientists, who are familiar with that particular developing country and who may very well retain some link with it even after having returned to their own developed country to pursue their ultimate professional career. Their original contacts with local students and technicians may well lead to increasing opportunity for the local technicians to study eventually in one of the major university centers abroad.

The youth and relative lack of formality of young postdoctoral fellows (frequently with no dependents or with small children who pose no schooling problems) is in my opinion an enormous advantage in the interaction of such a foreign group with local technicians, students, and apprentice research workers. Ultimately, as qualified research scientists are trained in that country, they will assume permanent positions in the research center, but the flow of foreign postdoctoral fellows should not be cut off, because it will be the most effective protection against scientific inbreeding. A separate and more permanent administrative staff would have to be provided, but given that its scientific level can be very low, no particular difficulty should be encountered in finding such personnel.

4. Operational Problems: The ability to attract the required part-time directors, who in turn will be responsible for attracting the postdoctorate fellows, is related to the quality of the research facilities. Although it is obvious that most advanced research is more easily performed in New York, London, or Zurich than for instance in Nairobi, it is indispensable that unnecessary complications be eliminated. Complete duty exemption and very rapid importation by air without delays in customs are minimum conditions, because numerous items for research will have to be brought in continually. First-class equipment, an appropriate scientific library, good physical facilities, and suitable and reasonably priced housing for the foreign postdoctoral staff are essential.

The research center may well have a formal connection or in fact be part of a university or government institution, but in any event there should be intimate informal contact with a university if one exists. In some locations (certain Latin American countries immediately come to mind) official separation from an existing university may be desirable because frequently such institutions are bound to have rules, salary structures, and bureaucratic procedures that could not be tolerated in the research center.

5. Financial Funding: The Mexican precedent of private funding by local industry is obviously not relevant, because, if the economic incentive were present, such a research center would already be in operation in the hypothetical country under consideration. A ten-year commitment is probably a minimum, and it is likely that the bulk of the funds would have to come from abroad.

It would be preferable if the funds did not come from foreign government sources or at least not from one single foreign country. A consortium of private foundations with possible contributions from some major private industries funneled through some international body might provide a feasible mechanism. My opinion is that the question of funding should not be considered until some planning on questions 1 through 4 had first been carried out and a more precise budget determined.

CONCLUSION

Many problems that I have not considered are associated with the establishment of such a research center. Nevertheless, the main question reduces itself to one of whether it is in fact worthwhile to establish one or more centers of excellence for basic research in a developing country that, according to traditional criteria, is not yet ready to perform such research with its own manpower and financial resources. In my opinion it is worthwhile to undertake such an experiment, especially when it is coupled with the type of part-time director and postdoctorate fellowship program outlined under Staffing. This may be the prototype of an international scientific "peace corps" of small numbers and very high educational caliber, which may prove viable and useful in other areas as well.

Note: Within two years of the publication of this article, a consortium of academies of science established the International Centre for Insect Physiology and Ecology (ICIPE) in Nairobi by following precisely the prescription outlined. ICIPE is now a completely Africanized research institution with a multimillion dollar annual budget.

15

Pugwash, Population Problems, and Centers of Excellence

The population problem can be reduced to two straightforward questions. First, how can we feed, shelter and, in general, provide adequately for the present population? Second, how can we reduce the enormously accelerating growth rate of the world's population? In view of the very long lasting numerical effects of uncontrolled fertility (even if zero population growth were instituted in the United States now, its population would stabilize itself only after 60 or more years because of the children already born), enormous attention must be paid to the second problem, because of the numerical burden[1] that is imposed on future generations with each passing year during which fertility control is not instituted. What can Pugwash contribute to this second problem?

A ROLE FOR PUGWASH

Various aspects of the world's population problem have appeared on the agenda of past Pugwash Conferences, notably in the Working Group dealing with science, technology, and development. Neverthe-

[1]At present (1993) global growth rates, we add annually the equivalent of one Mexico (the eleventh largest country in the world with circa 90 million people) to the world population.

Note: Original text written in 1970 and 1971.

Pugwash

On July 9, 1955, during the height of the cold war, Bertrand Russell issued the now famous Russell–Einstein Manifesto, cosigned by seven other Nobel laureates. "In the tragic situation which confronts humanity," it started,

> we feel that scientists should assemble in conference to appraise the perils that have arisen as a result of the development of weapons of mass destruction, and to discuss a resolution in the spirit of the appended draft. We are speaking on this occasion, not as members of this or that nation, continent or creed, but as human beings, members of the species Man, whose continued existence is in doubt. The world is full of conflict; and, overshadowing all minor conflicts, the titanic struggle between Communism and anti-Communism.

Four days later, Cyrus Eaton, a maverick capitalist, then chairman of the board of the Chesapeake & Ohio Railway Company, wrote to Russell:

> Your brilliant statement on nuclear warfare has made a dramatic worldwide impact. . . . Could I help toward the realization of your proposal by anonymously financing a meeting of the scientists of your group at Pugwash, Nova Scotia? I have dedicated a comfortably

less, the impact and involvement of Pugwash in this global problem has been vanishingly small. Two reasons come to mind immediately to explain this apparent neglect. One is the professional composition of Pugwash participants, which leans heavily on the physical sciences and is particularly strong in the disarmament area. The other is the veritable magnitude of the population problem, which raises the question whether Pugwash can make a meaningful contribution in this area.

One necessary, though admittedly not sufficient condition, for effective birth control is the availability of acceptable contraceptive procedures. Research in human reproductive physiology is extremely complicated, and the technical and financial resources for such research and for the ultimate development of practical new contraceptive agents are concentrated largely in the developed countries, notably in North America and Western Europe. A consequence of this situation is that virtually all of the newly introduced contraceptive agents (e.g., oral con-

> equipped residence there by the sea to scholarly group. . . . I suggest Pugwash because I believe you could more readily focus the attention of the world on the problems you wish to stress by meeting in such a relatively remote and quiet community than by choosing one of the great metropolises where the gathering would be but one of a number of events competing for public notice.

The first meeting was held in July 1957 in Eaton's family home and attended by 22 scientists (mostly physicists, with a sprinkling of chemist and biologists and one lawyer) from ten countries. The three largest delegations came from the United States, the Soviet Union, and Japan. There was no hotel in Pugwash, and the participants were put up in Eaton's house and in three railway sleeping cars. This modest beginning led to the annual Pugwash Conference on Science and World Affairs, which, though subsequently held around the world in places such as Kitzbühel, Moscow, Stowe, Dubrovnik, Udaipur, and Venice, continued to carry the name of the Nova Scotia village. Insiders refer to themselves as "Pugwashites," many of them persons who subsequently acquired real power in government. (Henry Kissinger, for instance, is a quintuple Pugwashite, as is Georgi Arbatov, for many years the Soviet government's top advisor on American affairs.)

traceptives) were developed with women of developed countries, or of a relatively high economic level, as the ultimate targets. The use of such agents in the developing countries and especially among very poor populations became an afterthought, and many epidemiological, cultural, and political factors were initially not taken into consideration. It is fashionable, among scientists as well as laymen, to state that better or more suitable contraceptive means or alternatives to existing methods should be developed, and long lists that might be considered from a theoretical scientific viewpoint have been prepared, but very little, if anything, has been done about *practical* and logistic proposals that are germane to poor populations of developing countries (e.g., Egyptian peasants in the Nile Delta, urban slum dwellers in Calcutta, and Peruvian peasants in the Altiplano, to name just a few extremes).

Under the "rules of the game" now operative in North America and Western Europe, development times for fertility control agents of the order of 10–15 years are optimistic estimates. This is a shocking conclusion if one remembers that, at current growth rates, the world's population is doubling in less than 30 years.[2] Let me ask again, what can Pugwash contribute to this problem? Its financial and technical resources, or potential, are so limited as to preclude any meaningful operational contributions. However, Pugwash can play an important catalytic role, because it is nonpolitical, nongovernmental and, most important, has no built-in vested interest in the birth control field.

In my opinion, one of the most inhibiting factors to more rapid scientific development of new contraceptive advances is the interplay (positive as well as negative) of family planning activities with scientific studies concerned with human fertility control. The *positive* interaction is required because no contraceptive procedure, however attractive and logical it may appear on scientific grounds, will be used on a massive scale unless it fits within the family planning limitations of the given country or population group. Social, religious, political, and economic factors are crucial components of a family planning program, and the scientist intending to work on novel fertility control procedures must take these factors into consideration at an early stage. Unfortunately, this has not been generally the case—especially because most of the scientific work has been carried out in the most advanced countries, while the practical applications to fertility control in less developed countries (LDCs) came much later and, generally, as an afterthought.

The *negative* interaction between family planning objectives and scientific research development can be insidious. Generally, governments jealously guard their prerogative to establish their own family planning objectives, and many sensitive political considerations frequently override cogent economic or environmental arguments.

Thus, everybody can list numerous countries all over the world that, on logical grounds, should have implemented major fertility control projects among their own populations but have not done so for political reasons. This situation has either led to a relatively low level of scientific research in reproductive biology and fertility control in those countries where the potential manpower is available (e.g., Russia and other Eastern European socialist countries or Japan), or else stifled the creation of an adequate pool of local scientists who are active in this field of research and are potentially capable of developing birth control agents suitable for their own populations (a situation existing in many LDCs).

[2]Based on 1992 rates, the doubling time is now 42 years.

This vicious circle, notably in the LDCs, of being dependent on foreign technical developments in the birth control field, because local capabilities are not available due to currently unfavorable family planning climates, must be broken. It must be broken because, ultimately, all countries with uncontrolled fertility will want to institute family planning programs and it would be highly desirable if, at the very least, indigenous scientific manpower were available in the field of human reproductive biology and were working on locally acceptable approaches. By the time such research matures into practical feasibility, the political climate, with respect to family planning attitudes, is bound to change.

Because of Pugwash's international and multidisciplinary composition and its independence from government agencies, I propose a "traveling Pugwash symposium" that might make an important contribution to the problem of creating new and more suitable contraceptive procedures for developing countries. Even more importantly, it might bridge the gap between the scientists (usually from the most highly developed countries) active in research in human reproductive biology and fertility control and the social scientists and politicians (from LDCs). The main purpose would be for "local" social scientists and politicians from a given LDC to define and describe a "perfect" contraceptive suitable to their cultural, political, and economic milieu, which a Pugwash group of scientists (representatives from disciplines such as medicinal chemistry, endocrinology, clinical medicine, and reproductive physiology) would then evaluate. This second group would pinpoint the areas of basic knowledge that might be missing and need to be developed, or conversely the group might call attention to already existing chemical or biological hints that could be exploited fairly quickly.

After several such Pugwash symposia have been completed at representative sites in Latin America, Africa, Asia, and the Middle East, a summarizing report would have to be written, which could be invaluable for U.N. agencies (e.g., the World Health Organization), World Bank, government regulatory bodies (e.g., the Food and Drug Administration in the United States), and the general scientific community. Such a report would indicate for the first time what crash programs (basic or applied) are necessary to accomplish an important objective—namely, to develop new contraceptive agents suited to important regions of developing countries and to reduce the enormous lead times that are currently required under traditional operating procedures of developed countries. Such a report would help in redirecting present research and development efforts in contraceptive technology, which do not take into consideration the peculiar epidemiological factors of a given cultural group. Pugwash could be an excellent vehicle for such a planning purpose, and the objective is sufficiently important and the "traveling symposium" approach sufficiently simple and realizable, that it merits attention.

AN AFRICAN CENTER OF EXCELLENCE IN REPRODUCTIVE BIOLOGY

As has been pointed out numerous times, the biggest bottleneck in scientific research of a fundamental nature in human reproductive biology, as well as in more applied work dealing with the development of practical fertility control agents, is the unavailability of suitable animal models. Reproduction is probably the most species-specific property, and only the higher apes appear to be suitable substitutes for humans. Indeed, extensive metabolic work with steroids in various primates has demonstrated remarkable differences among various monkeys, and between monkeys and humans. The international need for primate facilities is serious and, at the same time, so is the need to prevent extermination of apes. Theoretically, gorillas and chimpanzees might be the best experimental animals, but it is obvious that gorillas are hardly feasible from a practical standpoint. Even the chimpanzee is not the easiest animal to handle, chiefly because of its size. The research requirements, in terms of numbers, will be such that it is essential that adequate studies and facilities be made available for the breeding of chimpanzees; otherwise their eventual extermination is ensured. Insufficient efforts have been expended on the breeding of chimpanzees in captivity to solve this problem on a massive scale.

By far the greatest natural reservoir of gorillas and chimpanzees is Zaire and surrounding areas. This circumstance probably can be utilized to solve an international problem and, at the same time, contribute to the problem of training scientists in developing countries in disciplines that may be of particular relevance to them. During a trip (1971) to Zaire as the chairman of a U.S. National Academy of Sciences delegation to a United States–Zaire workshop on science and technology, I outlined some of these problems.

I propose the establishment, similar to the International Centre for Insect Physiology and Ecology (ICIPE) in East Africa, of an international center in the Congo region dealing with primate biology.[3] Organizationally, a consortium of international academies and scientific bodies would fit exceedingly well: just as with ICIPE, major scientists of international reputation could commit themselves to serve as part-time research directors through periodic short term visits, and especially through provision of longer term stays (two–three years) of young postdoctoral fellows from their laboratories. The function of these highly trained younger postdoctoral fellows would be to conduct research at the Primate Biology Center and, most importantly, to train Congolese and other Africans for research and development in areas germane to the center. Research in primate reproductive biology covers a wide spectrum of closely related fields—metabolic and physiological studies, toxicology, endocrinology, etc.—and may ultimately be extended to other medicinal and pharmaceutical studies using primates as main experimental objects. The attraction of such a center is its gradual establishment, proceeding slowly through several logical stages of development.

The first and indispensable stage would be the establishment of breeding colonies in a natural habitat to ensure an adequate supply of animals. Attention should center on the pygmy chimpanzee (bonobo), which appears to be fairly prevalent through the central Congo region but which, until now,[4] has never been used as an experimental animal. Its smaller size should make it more useful than the chimpanzee, and it may thus be possible to introduce a new species into experimental pharmacology and toxicology.

If this first stage succeeds—namely, the establishment and development of breeding colonies in the pygmy chimpanzee's natural habitat, preferably on an island—then the next one would be to expand the center to a research establishment starting with reproductive biology and extending gradually into many of the related disciplines. As the scientific scope of the center increases, so will the number of outside, part-time research directors, the number of postdoctoral fellows and, most

[3]The political situation in Zaire in the 1990s makes such a proposal totally impossible whereas 20 years earlier, when this concept was first launched, the local and international climate was much more receptive to such an idea.

[4]As a result of this proposal, four bonobos, accompanied by a Zairean veterinarian, traveled to the Yerkes Primate Center in Atlanta (for extensive details, *see* Chapter 14 in C. Djerassi, *The Pill, Pygmy Chimps, and Degas' Horse,* Basic Books, New York 1993). Substantial research over nearly 20 years has since shown that bonobos are probably the closest primate relatives of humans. In theory, they would thus be the ideal animal model in reproductive biology.

importantly, the number of African scientists. In fact, the first phase (establishment of breeding colonies) would require several years, so a select group of Africans could, in the meanwhile, be trained abroad in preparation for research at the Congolese center once it became operative in its second (research) phase.

If successful and, if operated along the ICIPE model, such an institution in Central Africa could become a significant research and training center in reproductive biology, without introducing the question of family planning which, at this stage, is anathema to many African countries. An enormous amount of necessary basic research that could be performed there would be of relevance to fertility control programs in non-African countries, and much work on biological problems of relevance to African countries could be accomplished. Furthermore, if, in the future, plans are instituted by African countries to develop novel fertility control procedures that may be particularly suited to their own circumstances, a center staffed by local scientists might be available with resources upon which those countries could draw.

Miscellaneous Topics

16 Insect Control: Operational and Policy Aspects

The original article, on which this revision is based, was co-authored with Christina Shih-Coleman and John Diekman.

What common factors make significant advances in insect and human population control so difficult? Much has been written about the economic importance of insect control, the relation between such control and the available food supply, the need to prevent environmental degradation, and the theoretical availability of various alternative approaches to insect control that could or should be developed. Let us assume that the desire for integrated pest management is not just pious preaching but actually represents a real goal of government policy makers, scientists, and eventual users. (Integrated pest management refers not to the abolition of chemical agents for insect control, but to the judicious use of such agents together with biological, biochemical, and microbial methods as well as the application of other control techniques.) When, if at all, can we expect the widespread availability of such new methods? What will it cost—especially in terms of time—to create them? What policy changes, if any, should be made in order to expedite or even make possible the practical realization of such goals?

In 1970 I analyzed some of the operational problems (cost, time, and regulatory barriers) and policy questions in the field of human fertility control that have to be considered in converting laboratory discoveries into practical methods suitable for millions of people. Hardly any attention seems to have been paid by the layman or the scientist to the many

Note: Original text written in 1974.

similarities that exist between insect and human fertility control in terms of either the operational aspects of such research or governmental policy in these areas. It is also interesting that in both human fertility control and insect control most of the important methods currently used are largely chemical in nature, and that attempts to develop new methods center on trying to develop nonchemical alternatives. Policy makers in both fields generally do not recognize how difficult it is to accomplish such changes and improvements.

SIMILARITIES BETWEEN HUMAN AND INSECT POPULATION CONTROL

1. The development and eventual commercialization of agents for human fertility and insect control is governed by regulatory agencies: the Food and Drug Administration (FDA) in the first instance and the Environmental Protection Agency (EPA) in the latter, although the FDA and the U.S. Department of Agriculture (USDA) may also participate in the regulation of insect control agents under certain circumstances. In contrast to the situation in the 1960s, these regulatory agencies not only monitor existing products on the market or decide on the release of new agents to the public but, much more importantly, they wield an enormous prospective effect on research and development (R&D).

2. Virtually all agents in practical use for human fertility or insect control have been brought to the marketing stage by industry rather than by academic, governmental, or nonprofit organizations, even though nonprofit organizations have frequently contributed heavily at the research level. This situation in free enterprise countries has several important consequences that are usually given scant or no attention by government policy makers. One such consequence is that return-on-investment (ROI) calculations become important, and if the developer (industry) does not see a potential market that is attractive enough to permit recovery of the research investment, research will not even be started in that area. In general only very large companies with substantial financial and manpower resources are active in these two fields.

3. There is little formal inquiry into the reluctance of industry to perform research in certain areas of fertility control, be it for human beings (e.g., male fertility agents) or insects (e.g., biological or biochemical methods). Policy makers ought to consider whether special incentives are needed to stimulate the conduct of practical work in these areas or whether other sectors of the R&D community could contribute effectively to practical new approaches in these fields. *I make the categoric*

prediction that, if this is not done promptly, most current public pronouncements on the likelihood of fundamentally novel human or insect population control agents will represent grossly optimistic exaggerations.

4. The great public clamor for improved agents is associated with an equally great clamor for absolute safety. Because nothing is absolutely safe, a statement about safety is purely an estimate of degree that should enable us to choose among alternatives. However, this choice is never based on scientific considerations alone, but also on political ones. The difficulties of making such risk–benefit decisions are especially serious in the fields of human and insect population control.

5. In spite of the virtually unquestioned necessity for improved agents for human and insect population control and in spite of the urgency of the situation, practically no financial support or other incentives are provided to industry by the U.S. government (in marked contrast to the situation existing in the 1970s in countries such as Germany and Japan) for R&D in the insect control field.

DIFFERENCES BETWEEN HUMAN AND INSECT POPULATION CONTROL

1. The public's concern for the environment is one of the factors motivating the development of new agents for both insect and human population control. However, such concern is really very different in the two areas. Contraceptives involve primarily the microenvironment of the individual and, furthermore, their use involves voluntary decisions by the consumer. Insect control has macroenvironmental consequences. Furthermore, with pesticides, the ultimate consumer of a crop that has been treated has usually not participated in making the decision to treat that crop. Thus it is even more difficult to make risk–benefit determinations in the field of insect control than it is in the field of human population control.

2. The scope for financial return on R&D investments is much more circumscribed in the insect control area than it is in the field of drugs, such as contraceptives, for human beings. Thus, if one develops a fundamentally new drug for human disease, the subsequent commercial price for this drug usually can generate a return that is commensurate with the original R&D expenditure, because a higher economic limit is placed upon cures of diseases or prevention of death than upon the control of insects. With insect control, even if one develops a fundamentally new and environmentally superb agent, its cost is immediately

controlled by the economics of the crop that one is going to protect. In other words, with the possible exception of certain public health pests, the economic cost–benefit is a prime determinant in insect control, whereas in human fertility control humane as well as human factors (cultural, religious, or political) intervene heavily.

3. Return on investment (ROI) calculations in the two fields depend on the respective potential overall markets, which are very different. Demonstration of clinical efficacy is essentially applicable anywhere in the world, and most human applications of drugs, including contraceptive ones, have potentially a global market. This is hardly true of new insect control agents, although it may have been the case at one time with DDT. The understandable insistence by public and also many scientific groups for nonpersistent and much more selective insect control agents implies that in the future there will be numerous small- or medium-sized markets for many different agents and that ROI considerations and market dollar volumes will, therefore, be even more crucial. Development costs are now so high that industry is reluctant to develop narrow-spectrum pesticides unless the pest species is of major economic importance. As development costs for new pesticides increase, the large companies that can best afford the expense will avoid small, specialized markets, whereas the small company that might be interested in the small market will no longer be able to afford to enter it. It follows that in the field of insect control, the progress and extent of R&D will be even more sensitive to the regulatory climate and to the length of development times than it will in the human field.

4. Another important difference between human and insect population control is that, in general, local factors play a much smaller role in human beings than they do in insects. In human beings, an effective cancer cure or a male contraceptive developed in London is almost certain to be effective also in San Francisco, California, or Johannesburg, South Africa. On the other hand, an effective control agent developed in one geographical area for a particular insect does not at all mean that the same insect will be controlled in another geographical setting because of differences in climatic conditions, agricultural practices, or the presence of other insects. It is therefore particularly important to conduct trials of such insect control agents in many different geographical locations. Toxicological tests, however, provide results that do not vary according to where the tests are conducted, and regulatory practices should be designed to facilitate the worldwide acceptance of such toxicological data. (Japan, for example, usually does not accept the

results of toxicological tests performed in other countries). Unless one deals within the legislative borders of a large geographical entity, international cooperation is required to avoid duplication and delay.

5. Regulatory agencies operationally determine when serious applied research can be started with human beings (when clinical research can be conducted under an "investigational new drug permit") or with insects (when substantial field trials can be conducted under an "experimental use permit"). This similarity, however, introduces one of the most significant differences in the two areas and one where the regulatory process can have a disastrous effect. With relatively few exceptions, it does not make a great deal of difference when the FDA permits the initiation of clinical work because studies on cancer or tuberculosis, for instance, can be started in January as well as in June. This is not at all the case in the field of insect control, where a delay of only a couple of months by a regulatory agency in acting on an application for an experimental use permit may cause a project delay of an entire year because most insect life cycles are restricted to relatively short seasons, except in tropical regions.

6. Once the newly developed agent has been approved for public use, there is a great difference in the manner by which it is disseminated. In human fertility control, the distribution system is under constant regulation and surveillance by members of the medical profession. In insect control—in actuality a more complex field—decisions are frequently made by lay persons.

NATURE OF INSECT CONTROL AGENTS

Given the right research climate, advances can be expected in the following four areas of insect control, which are subject to government regulatory scrutiny.

Chemical Insecticides. Most research, notably in industry, is concentrated on chemical insecticides, especially on modifications of existing structural types such as organophosphates, chlorinated hydrocarbons, and carbamates. The emphasis is on agents with less persistence, improved efficacy, and greater specificity to target organisms, but in the process a price is frequently being paid in terms of toxicity to humans or mammals. Thus the persistent DDT is less toxic than the nonpersistent organophosphate parathion. Both are relatively nonspecific agents—an economic advantage but an ecological hazard.

"Biorational" Chemical Agents. Another chemical approach, which is associated with a considerable amount of scientific glamour, can best be

categorized as the use of "biorational" chemical agents. We have coined this term in order to avoid the great confusion in the literature between chemical and biological control of insects. Typical examples of biorational chemical agents would be pheromones (that affect insect behavior), insect hormones (for instance, insect growth regulators), and hormone antagonists, all of which are frequently classified as biological control agents. A pheromone, which is an insect secretory product, is as much a chemical as DDT, but its mode of action is based on a completely different rationale. Thus the use of a pheromone, by its definition, usually involves a species-specific agent that is often active in very low concentrations and generally is neither persistent nor toxic.

Juvenile insect hormone mimics, or insect growth regulators, usually lack the extreme specificity of pheromones; therefore, their potential commercial applications are clearly wider. Their unique biochemical mode of action appears to limit their effect to members of the phylum Arthropoda while rendering them relatively innocuous to humans and other animals. In fact, some of them are the least toxic and least persistent chemical insect control agents known.

Microbial Agents. This is an area in which a substantial amount of academic and government research has been conducted and to which industry has also made some significant contributions. Two bacterial agents, *Bacillus thuringiensis* and *B. popilliae,* are used commercially. Also (and this is an indispensable first step toward the ultimate open use of such agents), an exemption from the requirement of a tolerance has been granted for the nucleopolyhedrosis virus (NPV) of *Heliothis zea* (the corn earworm).

Some experience has already been gained concerning the cost of developing microbial agents and the possible barriers to our putting them rapidly into use. Although few investigators believe that microbial agents will replace chemical methods, there is a general consensus that they may play an important role as supplementary agents in many integrated pest management programs.

Biological Control Procedures. Classically, the biological control of pest insects refers to the introduction of natural predators and parasites of accidentally imported pest species or, in some instances, of endemic pest species. In principle, biological control offers nontoxic, nonpolluting, relatively inexpensive, long-lasting, and self-perpetuating protection. The use of genetically modified insects, such as the use of sterile males in controlling screwworm flies, sometimes is erroneously called a biological control procedure. This technique does not constitute true biological control because new batches of sterilized males must be released periodically, and thus continual intervention is required—a feature that makes the technique more expensive. Furthermore, it is most likely to work when applied to very large areas or areas with natural geographical barriers (e.g., islands) where continuous outside infestation is prevented.

The use of natural enemies is a vital facet of integrated pest control. Unfortunately, biological control is not a panacea. First, the introduction of beneficial parasites and predators usually does not offer immediate control. Second, some level of insect damage must be expected because a certain minimal pest population must be maintained if the predators and parasites themselves are to survive. This unavoidable level of crop damage may not be acceptable economically to farmers or esthetically to buyers. Use of foreign beneficial insects against pests of domestic origin has been little explored because such an introduced natural enemy would not only be poorly adapted to the target pest but would have to contend with an unfamiliar and perhaps less hospitable environment and compete with already adapted domestic natural enemies of the pest species.

"Cultural" control practices for insects are also of potential value. Such practices include the adjusting of planting and harvesting dates to avoid insects, the use of plant varieties resistant to insects, the implementation of deadlines for destruction of crop residues at the end of the growing season, and the careful control of irrigation water to avoid the breeding of mosquitoes. All of these cultural practices can be improved by research and all are or can be subject to government regulations that may enhance or hinder their development and adoption. Research advances in cultural control practices are almost certain to come from the public sector.

REGULATORY REQUIREMENTS

In 1947 the Federal Insecticide, Fungicide, and Rodenticide Act (FIFRA) was passed. It called for USDA registration of economic poisons prior to their interstate sale or transport. Approval of registration applications was based upon review of safety and efficacy data. It also required that product labels contain instructions for use and warning statements to prevent injury to human beings, other animals, and plants. In 1954 the Miller Amendment to the Food, Drug, and Cosmetic Act required the FDA to establish tolerance limits for pesticide residues remaining on raw agricultural products by review of toxicological, metabolic, and persistence data. In 1970 full regulatory powers (that is, registration and establishment of tolerances) over economic poisons was transferred to the EPA.

It was felt that FIFRA had a number of weaknesses (including lack of direct federal control over pesticide use, lack of any enforcement powers other than lengthy, cumbersome cancellation procedures, and lack of authority to regulate intrastate manufacture and shipment of economic poisons), so in 1972 the Federal Environmental Pesticide Control Act (FEPCA), generally referred to as "FIFRA as amended," substantially changed FIFRA.

A pesticide is defined by FEPCA as "any substance or mixture of substances intended for preventing, destroying, repelling or mitigating any pest. . . . " This all-encompassing definition calls for EPA registration not only of chemical agents (traditional insecticides and biorational agents), but also of living organisms (microbial agents and conceivably even beneficial insect predators and parasites). In actual practice, the EPA does control microbial and biorational and traditional chemical agents, but developers and users of insect predators and parasites are regulated to a considerable extent by the USDA and some state agencies.

The USDA guidelines primarily call for assurances that the imported or transported insect will not pose a potential hazard to crops, human beings, other animals, or beneficial insects, and that no hyperparasites or other insect species are included in the shipment. Safeguards to prevent escape during shipment are required. Thus, the USDA requires assurances of safety, but not necessarily efficacy, in the use of beneficial insects as a means of insect control, unlike the EPA, which spells out detailed safety and efficacy requirements.

TIME AND COST ESTIMATES FOR THE DEVELOPMENT OF NEW AGENTS

Safety, efficacy, and environmental requirements for chemical insecticides are the standards by which practically every new insect control agent is judged by the EPA, even though such agents may bear little resemblance to classical insecticides. The cost of developing any new method of insect control falling within the scope of the EPA is going to be at least as expensive as the cost of developing a traditional chemical insecticide. In addition to the baseline requirements established by the chemical insecticides, new agents will be required to undergo unique and additional studies because of their novelty, thus increasing the development time. (In 1970, a survey of the pesticide industry estimated an average of 77 months for the development of a new chemical insecticide.)

Biorational Chemical Agents. Little, if any, attention is paid to the fact that these agents are much less toxic and much more species-specific than classical insecticides and, because they interfere with the insect's own communication mechanism, their use can be described as "biorational." Costs associated with synthesis, production, and formulation are considerably higher for biorational agents than those reported for traditional chemical insecticides. Biorational agents (e.g., pheromones and many insect juvenile hormone mimics) differ totally in chemical structure from standard organophosphates, carbamates, and chlorinated hydrocarbons; thus, existing synthetic procedures, pilot plants, and manufacturing facilities cannot be slightly and inexpensively altered to meet the demands of each new chemical. Instead, entire new processes, equipment, and facilities must be developed and constructed. The same is true of biological evaluation because the mode of action of biorational agents differs considerably from traditional agents, necessitating more manpower for the development of new techniques. Finally, because of their higher specificity, the market potential is bound to be more limited compared to that of a broad-spectrum, classical insecticide.

Microbial Agents. Because viral agents are expensive to produce, have a relatively narrow range of activity, and are essentially nonpatentable, extraordinary regulatory delays act as strong impediments to the introduction of such insect control agents. Much of the caution exercised by the regulatory agencies in passing on the first viral agent is probably understandable, but it is unlikely that the development cost can be recovered economically in a reasonable time. Unless some economic incentive for this type of work is created, it is doubtful whether much additional research will be performed by industry in the area of new viral control agents.

Biological Control (Beneficial Predators and Parasites). The development of beneficial insects as a method of insect control differs greatly from the other three approaches previously discussed. First, the regulatory agencies involved are the USDA and certain state agricultural departments, rather than the EPA. Second, the universities of California and Florida, the State of Hawaii, and to some extent the USDA are the leading groups engaged in the introduction and establishment of beneficial insects. The role of private industry is by and large limited to that of dissemination, by some small firms, of insects raised from cultures obtained from government or university sources.

The costs involved in the development phase of a biological control agent are very much lower than for the other agents discussed. The nature of the agent involved (insect as opposed to chemical) precludes the need for expensive facilities, such as sophisticated analytical instrumentation for residue determination or quality control. The capital investment in building an insectary is much less than that in building a pilot plant or factory for producing a chemical insecticide. Efficacy testing (that is, field testing) is very similar for all the agents. The difference lies in the required safety testing, residue tolerance determinations, and especially the administrative procedures involved in registration.

RECOMMENDATIONS

Almost all of the following policy recommendations are equally applicable to insect control and human fertility control. They are based on the premise that it is desirable to encourage continued or even increased involvement of the private sector in the field of insect control. Otherwise, steps should be taken to make the government control of R&D and production of new types of insect control agents more effective.

1. Research impact statements (*see* Chapter 19) should be prepared by regulatory agencies for the same reason that environmental impact statements are prepared to aid in technology assessment and societal control. In matters as critical as those under examination here, it is no longer reasonable simply to promulgate some rules and regulations and then to ignore their consequences. Regulatory personnel should be required to examine in a prospective fashion the impact of given regulations on future research—especially on that research that policy makers would like to see emphasized and to determine whether the introduction of such regulations would discourage or completely inhibit certain fields of research.

2. Access should be created to a funding mechanism of long-term toxicology, which might be the single greatest inducement for bringing smaller, entrepreneurial companies into a field dominated exclusively by corporate giants interested solely in giant markets. Frequently, the results of toxicological studies are required before extensive practical field trials are even feasible (e.g., in the control of insects in stored

grain) and, because in the search for specific insect control agents several such products would have to be examined, the cost very rapidly escalates to such an extent that some potentially promising products are never put to practical test. This is especially true if the potential market is relatively small. Alternatively, only one product at a time is tested, thus extending the development time by years. The high burden of such toxicological requirements is even felt by some of the giant companies in the field, and has led to the suggestion that the public should bear a part of the cost.

The developer of the potential insect control agent should have the option of applying to the appropriate government agency—ideally the EPA—to initially fund the long-term toxicological studies, which would then be conducted in an outside laboratory. The performance of these studies by a third party is desirable in any event to ensure completely unbiased interpretation. If the product eventually led to a commercial entity, then the manufacturer would be obligated to pay a royalty to the government agency, which would continue to be paid for the life of the product or until the entire costs (with interest) of the toxicological studies had been repaid.

Such advance funding by the taxpayer through the government agency should not restrict the proprietary position of the developer, because the latter would have already expended millions of dollars on chemical, biological, analytical, and acute toxicological studies before the stage is reached where long-term toxicological studies are justified and required. The taxpayer would simply be asked to share the risk of additional safety requirements so as to make return on investment calculations somewhat more attractive in a field where social goals (e.g., species specificity) automatically imply smaller markets and hence considerably lower return on the original R&D investment. If a product actually reached the commercial stage, then the taxpayer would be repaid for this advance. To put a possible qualitative as well as quantitative ceiling on such requests for partial government funding of toxicology studies, a peer review system (excluding industrial participants) would have to be instituted.

3. The experimental permit phase should be expedited. It is easy to demonstrate efficacy in the insectary laboratory, but it is a completely different matter to demonstrate such efficacy under actual field conditions that frequently involve large areas (hundreds of hectares) in order to make the results meaningful. Government permission has to be secured for such experimental permits and the required regulatory procedures introduce two kinds of delays. First, insects are generally active

during very limited periods of time, and sometimes a delay of just a couple of months in securing an experimental permit may lead to a year's delay in actual work. Second, residues of an agent may remain either on a crop or in an animal's meat or milk and, consequently, great caution is exercised both by the regulatory agency and the developer before field trials are attempted because under certain circumstances that crop or those animals may have to be destroyed. In many instances this has led to hypercaution and inordinate delays before serious testing could be started. The granting of such experimental permits could be expedited if two conditions are met.

Instead of relying on reports made by the developer after the completion of field trials, the regulatory agency should actually assign a member of its own staff to participate in the monitoring of the experimental permit work. This approach would have the enormous advantage of exposing regulatory personnel to the actual problems encountered in field testing and of making them cognizant of the many practical problems that arise during the development of practical insect control agents, particularly of the fundamentally new kinds of agents.

Funds should be made available, possibly through some type of experimental crop insurance program, to ensure the developer and the owner of the crop or animals to be used for the testing that they could carry out experiments at an earlier date without an inordinate risk of having to pay for the destruction of a crop or group of animals because they contained unacceptable residues. The actual cost of such an insurance system would probably be negligible compared with the social benefits of expediting field research concerned with novel insect control agents.

The preceding three recommendations are designed to expedite and facilitate the R&D of new approaches to insect control. The following two refer to incentives once the product has reached the granting of a full registration by the EPA and the stage of actual public use.

4. The regulations for patent protection should be modified. When one deals with long development times, frequently one-third to even two-thirds of the patent protection time is actually consumed by the development phase, and frequently the inventor or developer of the product who bore the brunt of the intellectual and financial gamble has only a few years left of proprietary patent protection. For agents actively under regulatory review, the 17-year patent lifetime should be reduced to ten years, but the ten-year clock should start running only from the day that a full registration is issued.

5. Incentive bonuses should be paid to farmers for the first year or two to encourage the use of environmentally desirable agents. The social goals associated with our ever-increasing concern for the environment require that an insect control agent be relatively nonpersistent, specific to certain harmful insects, relatively harmless to beneficial ones, and, of course, harmless to humans, wildlife, and domestic animals. Nonpersistence generally entails more frequent administration, and specificity to certain harmful insects frequently means "biorationality," which in turn often implies administration at a very specific time of the insect's life cycle. In any event, any such new agent—be it a biorational chemical one, a microbial one, or biological one—may require that substantial education be given to the applicator and the user. In a number of instances such personnel may have to be completely retrained.

All of these features, whether operational or educational in nature, must be included in calculating the cost of insect control. There seems to be no evidence that government agencies are aware that the payment of subsidies to users, as an incentive for them to employ environmentally more desirable but also more expensive techniques, would be better than allowing the economics of the market to dictate whether or not new agents are developed.

SUMMARY

Human and insect population control have several features in common, all of them indicating that the lag times in converting laboratory discoveries into novel practical agents (responsive to societal and environmental concerns) are increasing greatly and that return on investment calculations are becoming more and more significant in decisions related to the development of new agents. Such calculations are particularly important in the field of insect control because, by being more specific, the agents of the future are likely to cover smaller markets. Several recommendations for stimulating the development of new methods of insect control are proposed that are addressed primarily to policy makers. If they are not implemented, then these suggestions should at least stimulate others to make alternative proposals. If neither event occurs, then it is unlikely that there will be any fundamentally new approaches to practical insect control in this decade.

17

Pesticide Development—Sociological and Etiological Background

Pesticides are primarily used in two areas, public health programs and agriculture. However, the particular geoeconomic setting greatly affects the use pattern and indeed the overall attitude toward pesticides. For instance, in the pesticide and drug fields, most people high on the socioeconomic cultural scale pay more attention to the inherent drawbacks than the potential benefits of these agents.

The public health problems in the Third World are enormous and frequently associated with the insects and related pests that thrive in the more tropical countries. A great deal of lip service is paid to the control of diseases such as malaria, Chagas disease, and river blindness. In actual fact, the affluent part of the world is unwilling to pay much for the control of these disease vectors, either in terms of money or technological resource allocation, and the poorer countries simply cannot afford either. In the affluent areas of the world, the public health aspects of pest control are frequently dominated by recreational or even aesthetic considerations to the extent that a few flies or mosquitoes in an affluent suburban home can become targets of substantial financial expenditure. On a "per insect" basis, the allocations of resources are grotesquely skewed when one compares the wealthy and the poor.

The agricultural sector covers the entire range from intense, expensive, preventive pesticide usage in the high-technology agricultural

Note: Original text written in 1976 and 1977.

countries (most noteworthy the United States) to desperate and insufficient use for survival of one-crop agricultural countries. The developed countries are the sites of technological wealth, where all new developments and most manufacture of pesticides occur, but they are also the countries where the greatest concern is expressed (primarily by the nonuser public) about environmental effects. The Third World countries are totally dependent on the highly developed countries for present and future pesticides, but their own priorities or concerns are generally not taken into consideration for obvious economic reasons.

The standards of the "nonuser" environmentalist are generally based on long-term consequences without major consideration for short-term economic penalties. The user, on the other hand, concentrates on economic costs at the expense of long-term considerations. The governmental regulatory agencies—primarily because of the pressure from the nonuser community—are more concerned with safety and quasi-environmental factors; economic benefits or payback assume a very secondary role. Furthermore, the entire regulatory process in this country and in most of the other technologically advanced countries active in pesticide development (e.g. Germany, Holland, and Japan) by its very nature unavoidably penalizes the innovator.

From a longer term sociological standpoint, these conflicting priorities are not necessarily bad, because stresses of this type frequently produce desirable, long-term changes in society. Unfortunately, these diverging priorities coincide with major food shortages in areas such as Africa, which are unlikely to subside for decades. The only real short-term answer to this problem is not increased food production—especially in the absence of additional arable land—but rather decreased food and agricultural losses. This will require global rather than parochial outlooks by government agencies and demand decisions that need to be made over relatively short time frames. Both of these requirements are contrary to the modus operandi of government agencies or worldwide organizations. Unfortunately, rapid development of fundamentally new pesticides will require incentives—primarily operational rather than financial ones—that neither government agencies nor the general public may find too palatable.

18

Planned Parenthood for Pets?

The original article, on which this revision is based, was co-authored with Andrew Israel and Wolfgang Jöchle.

Concern about the human population explosion in the 1960s led to greatly increased local, national, and international efforts to improve human fertility control and to extend effective family planning in the 1970s. However, the problem of a population explosion among dogs and cats had received little attention from the scientific community. In 1973 we tried to show that the numerical increase in these animals bore many analogies to the human population problem, including some of the barriers that preclude a rapid solution.

The magnitude of the human population explosion had been recognized as soon as relevant demographic statistics became available as a consequence of the introduction of the human census in most parts of the world. The first dilemma faced by the investigator examining the dog and cat population is the poor quality of numerical data even in those countries with an advanced human census. For this reason, we recommended the inclusion of relevant questions on dogs and cats in the next human census in the United States.

For the United States, in 1971 there was a range of 20–90 million animals, according to various surveys—by far the highest number of such pets in the world on either an absolute or per capita basis. Put into other terms, while approximately 415 human beings were being born each hour in the United States, 2000 to 3500 dogs and cats were born

Note: Original text written in 1973.

(assuming an average lifespan of three to five years). Early puberty, large litter size, and shorter pregnancies make dogs about 15 times and cats 30 to 45 times as prolific as humans. The relatively wide range of the U.S. dog and cat population estimate was due to the decentralized and even disorganized nature of our animal control practices (facilities run directly by city, county, or state governments; private humane organizations contracted to a municipality; or private humane organizations operating independently), so that statistics often were not available on a county, let alone state, level. Registrations were a poor source for estimating total animal populations; less than half of all dog owners registered their animals, and most areas had no compulsory cat registration. Feral dogs and cats also pose a problem in the United States that is not seen in other countries such as the United Kingdom and West Germany.

ECONOMICS AND ECOLOGICAL IMPACT

The benefits of dogs and cats cannot really be quantified because many of them are of a personal and nonmaterial nature. This condition is not true of many of the costs.

The simplest dollar expenditure to calculate is that associated with the operation of a pound or shelter. Often run by humane societies, these places can serve as dumping grounds for unwanted pets. Surveys on both coasts indicated an average handling cost of $7.00 per animal in 1972—about $125 million annually. About 75% of this sum was consumed in killing unwanted or stray animals.

The disposal of carcasses of destroyed animals creates an environmental burden. Tons of dead animals are buried in city dumps, incinerated, or sent to rendering plants to be cooked, ground, and used in fertilizer or cattle feed. Where large quantities of dogs and cats must be disposed of, the solid waste and air pollution problems should not be overlooked.

The costs from cattle losses and wild dog control run to millions of dollars annually. Costs incurred with rabies control, dog bite care, sanitation, and public health care related to dog- and cat-borne diseases are also significant. Human infections, which may be acquired by pets and passed back to humans, are the human types of mumps, measles, tuberculosis, diphtheria, and scarlet fever. Leptospirosis (spread by dog urine in the soil or water supply), toxoplasmosis (spread by cat feces in houses), and cat scratch fever have assumed progressively more importance.

Urine and feces of dogs and cats present a particularly serious problem to young children who play in or eat dirt where infected feces or urine is spread. (Nationally, *daily* dog feces production amounts to about 3,500 tons.)

The aforementioned costs may be classified as involuntary ones. Voluntary pet expenditures in the United States are enormous as well. The pet food market alone reached $1.35 billion in 1971, and 6 billion pounds of dog and cat food were consumed.

For each dollar spent on food, dog and cat owners spend at least an equal amount on other products and services such as collars, clothes, deodorants, lotions, vitamins, beds, doggy and kitty hotels, dog psychiatrists, and even funerals. In addition, approximately $1.7 billion was spent for the purchase, licensing, inoculation, and veterinary care of dogs.

OVERPOPULATION

In spite of the huge sums of money spent on the care and feeding of pets, owners pay little attention to or are unaware of the number of surplus animals created by overbreeding. As a reference point, over a seven-year period the progeny of one unspayed bitch can range from 72 (controlled conditions) to nearly 4400 offspring under noncontrolled optimal conditions. American Kennel Club registrations exceeded 1 million in 1971, and pedigreed pet sales are definitely big business supported, in part, by at least 100,000 amateur dog breeders. Also contributing to the oversupply of pets is the family that feels a litter of puppies or kittens will be a "good experience for the kids."

The only approaches to a lower population, be it human or animal, consist of an increase in the death rate and/or a decrease in the birth rate. Litter destruction or intensive programs of euthanasia will raise the death rate, and this approach has been used more extensively than any other. Reduction in the birth rate is accomplished by vigilant owners confining their animals during heat periods or else subjecting them to surgical sterilization. The focus of surgical sterilizations must be the mature female dog (bitch) and the mature female cat (queen). Castration of the male dog and tom cat may reduce wandering and undesirable sexual behavior patterns, but it will not lower the birth rate while other fertile males are available for mating.

It is unlikely that much of the lay or scientific public is aware of the magnitude of the annual pet destruction in the United States, and the fact that this has been the principal method of keeping the burgeoning dog and cat population from becoming uncontrollable. Many millions of dogs and cats—some 12% of the total population—are destroyed each year at private and public shelters. California, the state with the largest canine population, reported the destruction of 435,237 dogs in 1970, a 7.7% increase over 1969. At least as many cats are destroyed, so the total California figure is in the range of one million animals per year. The majority of these animals are not injured, old, or diseased but are healthy pets relinquished or abandoned by their owners. In addition, many newborn animals are destroyed by their owners, and other animals are killed in traffic.

The 12% destruction rate in the United States should be contrasted with 1971 data reporting destruction rates of 2.4% for West Berlin and 1.6% for the United Kingdom. Japan is claimed to destroy few animals in pounds because the majority of stray dogs and cats are sold for use as laboratory animals.

Consistent with the widespread lack of knowledge concerning the extent of animal destruction is the fact that most states have laws punishing cruelty to animals, but few regulate animal destruction in pounds and shelters. Only in 1972 did California pass legislation regulating animal destruction by the three most common methods (carbon monoxide poisoning, high-altitude decompressions, and barbiturate administration) and providing for surveillance of such facilities to ensure humane practices.

PSYCHOLOGY OF PET OWNERSHIP

Ironically, Americans may have created the cruel pattern of dog and cat overpopulation and destruction in a quest to satisfy some of their own psychological needs. A substantial literature has developed about this topic, and only a few key conclusions are listed here to support the thesis that the psychological needs for pet ownership are also responsible in part for the opposition to more extensive surgical sterilization.

The American child's first exposure to animals is the anthropomorphized creatures frequenting juvenile fiction, thus establishing the humanlike qualities of pets at an early age. It has been suggested that pet ownership provides the crucial link between the animate and inanimate worlds and that pets often serve as scapegoats for their owners through mechanisms of identification, projection, and displacement. Other aspects, including the need for companionship to offset feelings of loneliness and inadequacy in the urban world coupled with a breakdown in human communication, have also been cited in support of the psychological need for pets.

Psychological considerations may explain why so many owners do not want their animals castrated. Many see puppies and kittens as substitutes for children not yet planned; or for the small number of children planned; or for children that never arrived or that have already departed; or as an unconscious protest against their own exercise in family planning. Some male owners may want their pet roaming and impregnating as an unconscious protest against the sexual restrictions society and morality impose upon them. Many people regard their pets as family members and are horrified at the concept of "taking sex away." For this reason, reversible chemical sterilants may gain wide owner acceptance, especially since they would have the additional feature of being more humane as well as more human in their application.

SURGICAL STERILIZATION

Although increase of the death rate in animals hardly has a human counterpart (except in major wars) and whereas advances in human birth control have had very little impact on pets, surgical sterilization is reasonably common in both humans and animals. Indeed, next to euthanasia, it is the most common method of controlling the pet population growth rate. Ovariohysterectomy, or "spaying" of the bitch or queen, involves the removal of the uterus and ovaries, thus rendering the animal sterile and stopping the estrous cycle. Common intrauterine diseases are virtually eliminated by this operation. Despite the health advantages associated with castration of the bitch or queen, the owner often wishes her to remain sexually intact, irrespective of whether she is to be bred.

The majority of pet owners can afford to have their pets spayed, but they are not always willing to spend the money. In an effort to reach those who cannot or will not pay private veterinarians their full fee for spaying, cities around the nation have set up low-cost spay and neuter clinics. The largest such facility was established in 1971 under the auspices of the Department of Animal Regulation of the City of Los Angeles, and the clinic ran only a small deficit the first year. The success and probable eventual self-support of this pilot program prompted the City Council to appropriate another $248,000 for extension of this program in 1972.

Veterinarians have frequently opposed public spay and neuter programs on the grounds that public taxes should not subsidize private pet owners (an argument that ignores the much larger indirect costs the public pays by not subsidizing such programs), that municipalities in supporting such programs infringe on private enterprise, and that such clinics are the first step toward socialized veterinary medicine. Nevertheless, the American Veterinary Medical Association set up a commission to study the pet population problem and to issue recommendations, and the California Veterinary Medical Association developed a platform on pet population, one plank of which recommended that low-cost spaying and neutering be done in existing veterinary hospitals rather than through the establishment of public clinics.

California law prohibits the release of any cat from a public pound or shelter unless provisions are made for the animal's castration. The owner is usually required to leave a deposit, which is forwarded to the veterinarian upon certification that the animal has been sterilized. The focus of many pound and shelter efforts lies in drying up the source of unwanted births by not releasing sexually intact animals.

It is revealing that by "surgical sterilization," nearly exclusively spaying of female animals is understood. Neutering of male dogs has never been seriously considered on a massive scale; and neutering of tom cats is predominantly wanted for the elimination of the smelly spraying (in-house territorial marking) and of the frequent abscess formations (resulting from fighting) which often makes veterinary intervention necessary. Other surgical sterilization techniques, such as tubal ligation and vasectomy, have rarely or never been seriously debated. Although highly effective as population control means, such surgical interventions would not eliminate nuisances like heat periods, roaming, love fights and plays, cohabitations, false pregnancies and pseudo-postpartum syndromes with nesting behavior, lactation, and dangers of mastitis and pyometra (accumulation of pus in the uterus, a life-threatening condition of older bitches and queens). Intrauterine devices (IUDs) would also not eliminate these nuisances or dangers. Anatomical difficulties with their insertion and the fear of pyometra induction have prevented their exploration as anticonceptive means in the bitch and queen. People wishing to have the problem resolved by surgery are probably more motivated by forestalling the nuisances of reproductive behavior than by the curbing of the population explosion.

Although advocates of public spay and neuter clinics as well as their opponents agree that each animal spayed is a step in the right direction, both groups admit that spaying is only a temporary measure for animal population control because veterinary manpower is insufficient for all the theoretically required spaying without seriously jeopardizing other veterinary services. The questions of veterinary manpower, cost, and owner motivation can most likely be answered by developing chemical birth control agents for dogs and cats.

No chemical birth control method can or should be developed without solid knowledge of the patterns of basic reproductive physiology in the target species. But the knowledge of reproductive physiology in the bitch and queen was, until the late 1960s, scanty and unbelievably incomplete. Despite their thousands of years of companionship with people, dogs were in this respect almost an unexplored field of knowledge. Sporadic progress in small areas and adaptation of concepts observed in other species, or acceptance of what has been seen in the dog and cat as valid for other species, has led in the past to enlightenment as well as to misconceptions—sometimes with disastrous results for animals and people. For instance, in 1853, the German scientist T. H. Bischoff discovered that bitches ovulated during a period of bloody vaginal discharge; he concluded that human females also ovulate during menstruation—a misconception that delayed progress in human reproductive physiology and gynecology for at least 75 years. In 1971 the use of the beagle bitch as a supposedly meaningful model for human response to progestins forced the removal from general distribution of several oral contraceptives.

The daily application of estrogen–progestin pills to bitches and queens, as used in human contraception, prevents ovulation as expected. However, such hormones stimulate a readily responsive and genetically conditioned endometrium to go all the way to a full maternal decidua (maternal part of the placenta) with no chance to shed this endometrium after a rhythmical short-time cessation of drug administration, as happens in human application. Even veterinarians have overlooked the fact that dogs cannot menstruate and have no chance to get rid of these overstimulated endometria, which subsequently convert to pyometra and make spaying mandatory.

Do dogs have an estrous cycle? Their wild ancestors all are animals with a single, seasonally well-defined heat period. As a consequence of domestication, the frequency of heat periods has increased in the bitch to estrous intervals ranging from 26 to 37 weeks, with well-recorded differences between breeds and a tendency to have more heat presentations during spring than during the remaining year, except for a second small maximum during late summer. In the early 1970s the existence of a true estrous cycle was demonstrated in the bitch in contrast to the previously assumed sporadic appearance of heat with long dormant periods of the whole system in between.

The queen's reproductive cycle is an anovulatory cycle. Ovulation in cats results only from coitus or mechanical cervix stimulation. If no mating occurs, the ripe follicles degenerate and shrink and the receptivity for mating (standing heat) vanishes temporarily, only to be resumed seven to ten days after a new set of follicles begins to ripen. Heat occurs at 20- to 30-day intervals during the spring and summer north and south of the tropics and throughout the year in the tropics. After sterile mating, the follicles rupture and form active corpora lutea, creating a pseudo-pregnancy. Cats have a postpartum heat (and at this time probably ovulate spontaneously) within 18 to 36 hours of giving birth and thus have an excellent chance to conceive again almost immediately. It is not surprising therefore that the quantitative consequences of completely uncontrolled breeding in cats staggers the imagination.

DRUG DEVELOPMENT AND ADMINISTRATION

Theoretically, several avenues are available for chemical contraception in bitches and queens. A postcoital agent is not desirable in the queen because the owner must have knowledge of the animal's mating and then administer a drug to prevent nidation (implantation of fertilized egg in uterus) or to abort the fetus. In the dog, postcoital estrogen administration is probably the most widely used birth control means, despite its risk of inducing endometrial disturbances leading to pyometra and the toxicity of exogenous estrogens in this species. An agent could be developed for the suppression of heat symptoms and ovulation thereafter, but this requires the owner to recognize heat—an enormous drawback. The major emphasis on drug delivery should be placed on a continuously administered agent that can be given either daily by the pet owner or applied once or twice per year by the veterinarian and that is initiated at a set and safe time during the bitch's cycle.

Long-term administration can be achieved through systemic administration by injections, implants, and pellets, or through incorporation in food. If put in food, the dosage range of the drug would vary widely with the animal's consumption. Even specially marked packages of food would not avoid accidental human consumption of contraceptive-containing food (possibly as much as 25% of pet food is consumed by humans). Regulatory agencies are quite clearly discouraging this approach. Biscuits (chewable tablets) dispensed by the owner and immediately swallowed by the animal are a superior approach. Ideally, their taste should repel children.

PREVENTION OF HEAT

Prevention of heat is analogous to the use of oral contraceptives in humans to prevent ovulation. But whereas the human's estrogen–progestin combinations are used for ovulation control, progestins alone can be applied in animals for the dual purpose of prevention of ovulation and of the behavioral and clinical heat symptoms, and for the avoidance of endometrial stimulation (anti-estrogen effect).

Prevention of nidation is analogous to a postcoital pill in humans. In bitches, the progestin megestrol acetate given orally during the entire heat season shows complete prevention of pregnancy without suppressing libido or heat symptoms. Bitches may also be given estrogens, 24 to 48 hours after misalliance to prevent nidation, although the subsequent risk of pyometra is increased. Nidation can be prevented in queens by oral administration of certain progestins within 24 hours after mating. The next heat occurs 25 to 30 days later.

The use of embryotoxic or fetocidal drugs would be analogous to the use of abortifacient agents in humans. Practically no work has been done in this area in dogs and in cats, and owner attitudes probably make the home administration of abortifacients unacceptable (e.g., abortion on the kitchen floor).

SEXLESS PETS

The creation of "sexless pets" seems theoretically possible in the light of observations made in rodents where injections of testosterone or estrogens in newborns prevented sexual maturation. No information is available on whether the same treatment would render dogs and cats permanently asexual. Delay of puberty may be possible; prepubertal treatment with a progestin effectively delays puberty for at least six months in the bitch to 20 months in queens without impairing subsequent fertility. Immunization with nonspecific but highly purified luteinizing hormone (of ovine or bovine origin) before the onset of puberty, as judged from results reported in rats and in rabbits might be a successful method to reversibly (or irreversibly delay puberty in the dog).

RECOMMENDATIONS

An effective program to control surplus dogs and cats must focus on many channels and must be implemented from the local to the national level. We recommend the following strategies.

- As a short-term answer, wider encouragement of surgical sterilization of dogs and cats with emphasis on the prevention of unwanted litters and the health benefits for the individual animal and its owner.
- Establishment and enforcement of higher licensing fees and especially stricter leash laws, or enforcement of existing ones, which would encourage more responsible pet ownership. The resulting funds should be made available to animal control agencies.
- Implementation of a tax on dog and cat food to provide funds that could be given to animal control agencies for educational programs and to research institutions for development of chemical sterilants and other contraceptive agents. Present research and development efforts in this area are grossly deficient and have hardly received any stimulation from government sources.
- Inclusion of pets in every census so that accurate and up-to-date pet statistics can become available.
- Encouragement of the surrender of unwanted animals rather than their abandonment.
- Encouragement of the ownership of smaller dogs. The trend toward bigger animals for companionship and protection purposes can contribute to a faster pet population growth.

19 Research Impact Statements

There is little question that environmental impact statements should and do play an important role in technology assessment and societal control. Would it not also be reasonable to ask that research impact statements be prepared by regulatory agencies? During the past few decades, the Food and Drug Administration (FDA) and, more recently, the Environmental Protection Agency (EPA) have assumed—either indirectly or by actual legal mandate—in certain fields the dominant role in deciding what research could or could not be done and especially how long it would take to bring such work to a decision point. Should the effect of such actions on research also be evaluated?

Research impact statements could be prepared either as internal agency documents or as part of an open dossier. My recommendation is that these documents be used primarily within the agency at first, in order to permit it to determine for itself what information and policies could be derived from such statements. Even such limited use would impose upon the staffs of regulatory agencies a mental discipline that is mostly lacking in the current decision-making process. Eventually, depending on the experience gained, the statements could become a regular feature, generally available and subject to refutation.

A typical research impact statement ought to include an evaluation (even if only a subjective one) of the research area that would be af-

Note: Original text written in 1973.

fected by given regulatory requirements. Major items that should be taken into consideration are the novelty of the research, the effects of the regulatory requirement on other areas, and, most important, a cost–benefit determination. For example, a given regulatory requirement might achieve a relatively minor gain in safety information at the expense of an important line of research. If so, what alternatives might provide such safety information without a substantial negative impact on research? What is the price in lost benefits that the public will pay through a considerable delay in the completion or total abandonment of a given project? The pharmaceutical field appears to be replete with such examples, notably in the field of birth control, and various people have claimed that the drastic reduction in the introduction of significant new drugs during the decades of the 1970s and 1980s was associated to a considerable extent with FDA-imposed requirements. If research impact statements had been required of the FDA during that decade, their review at this time and comparison with the actual research conducted would have been very useful in confirming or rejecting such claims.

Research impact statements would also be useful in encouraging truly novel approaches to pest control. Before substantial field trials with new pesticides can be undertaken, the sponsor of such trials must receive from the EPA an "experimental permit." Refusal of such permits usually prevents further development and presumably is based on real or hypothetical environmental considerations. Would it not also be desirable for these considerations to be accompanied by a statement that would evaluate the potential damage (that is, failure to replace presently used, persistent pesticides) if such research were *not* done?

The impact of regulatory agencies on research is now so enormous that they should bear some of the responsibility for *prospective* research planning—especially if the effect can be felt on a national scale. The research impact statement may be a useful device in calling attention at an early stage to the need for modification or even elimination of counterproductive regulatory practices.

20 "My Mom, the Professor"

Why isn't this phrase heard more often in the technologically most advanced country in the world, especially in the laboratory sciences like chemistry or physics? Is it a peculiarity of the discipline? Is it because there are so few women chemistry and physics professors in the upper ranks of academia? Or is it also due to the time demands of responsible motherhood and the 60- to 80-hour "macho" work weeks required of males and females alike during their pretenure life? Is it the ticking of the biological clock as the young woman receives her Ph.D. or M.D. degree around the age of 27, completes her postdoctorate stint near her 30th birthday, and enters, as assistant professor, the six-year race toward academic tenure in competition with her male colleagues? When should these superwomen decide to become supermoms?

Yet there are countries in which there are many more women (and mothers) among the higher academic ranks of scientists than in the United States. Argentina and the Philippines are two examples. The reasons for these differences are complicated, but one of them is the availability and affordability of domestic help, which permits raising an infant at home rather than having to depend on institutionalized child care.

Now that the American stigma of the working mother is rapidly disappearing; now that it is recognized that women are the largest untapped

Note: Original text written in 1988.

human resource in science; and now that more women graduate students are entering scientific disciplines from which they were earlier barred by cultural or operational factors, is it not time to take steps that would facilitate their decisions about childbearing and rearing? Let me offer one modest proposal along those lines.

These days, the bright young woman Ph.D. or M.D. has no difficulties securing fellowship support. In the majority of American universities, she can now compete openly for entering assistant professorship positions. What most cannot afford during that period is raising a child. Why not make available—on a competitive basis related to professional promise or performance—five-year grants (at a level of about $20,000 to $25,000 per year) for domestic child-care support? A woman scientist would be eligible to apply as soon as she has secured a postdoctoral or junior academic position, but actual payment and start of the five-year grant would commence only a couple of months before the expected birth of the baby and would be terminated if she resigned from her professional position. Would such a program stimulate some promising young women scientists to become mothers at a time when they would otherwise feel they could not afford it? Would such financial support attract some women into demanding scientific careers when they are otherwise not prepared to do so because of their desire for childbearing and child care in the home?

I propose a pilot program on the order of $1 million whereby a foundation or government agency would initially commit itself to fund perhaps ten such five-year grants. It would signal to American professional women that childbearing is not considered a biological burden but rather a societal benefit deserving societal support. The number of actual applicants will indicate whether such a scheme fulfills an unsatisfied need. At the end of the trial period, or perhaps when the majority of initial grantees have passed beyond three years of support, the recipients will be asked to report to what extent such a program has actually facilitated their decision to become mothers at an earlier age or to have children at all. If successful, such a program could then be enlarged and be made a permanent component of our science grant programs. It might even encompass other disciplines where the time demands of the profession and the obligatory absence from the home have also proved to be impediments to motherhood. "My mom, the professor" might then be heard more frequently.

21

Illuminating Scientific Facts through Fiction

The need for the popularization of science is almost as old as science itself. Like every generation with its own set of societal problems, ours thinks of today's problems as particularly acute. Current examples are the explosive growth of scientific information at a time when general scientific illiteracy is growing alarmingly; the complexity of "technological fixes" presented to a risk-aversive public suffering from chemophobia and oncophobia; the almost pathetic desire of legislators and communications media for simple black-and-white answers to intrinsically gray questions.

A related problem is the gradual change in the general public's perception of the character of scientists and the conduct of scientific research. The most glaring example is the current attempt of legislators—prompted by some recent, highly publicized instances of presumed fraud in research—to equate uncertainty and honest error with such social deviance, and to display the typical American knee-jerk response of "sue the bastards." I am concerned that in addition to not explaining properly the science we do, we are even less efficient in explaining the behavior and cultures of scientists.

To cite just a single gray example: most academics in the humanities are shocked by the multiple authorship practices of laboratory scientists, most notably the inclusion of authors—usually senior professors

Note: Original text written in 1990.

or laboratory directors—who have done none of the actual experimental work; many have not even participated in the writing of the manuscript. If professors of English do not comprehend that practice of their colleagues in chemistry or physics, how can we expect the great mass of scientifically unwashed legislators, let alone the general public, to have an understanding of, or sympathy for, our modus operandi?

As my small personal contribution to a better understanding of some of the unwritten rules of scientific behavior, I have chosen an infrequently used literary genre: science-*in*-fiction, in which the plot is populated by real scientists doing plausible science and the events occur in realistic settings. The initial responses by nonscientists to my experimentation with this literary form have led me to encourage others to do likewise. Important current issues—such as the underrepresentation of women in the higher echelons of most laboratory disciplines, the nature of the mentor–disciple relationship, the ethical conduct of research—have all been the subjects of many serious articles and books. But how does one get these writings before that huge segment of the reading public that never even sees such publications? Dressing these ideas in the garments of fiction is one way of introducing them to a new constituency. "Publications, priorities, the order of the authors, the choice of the journal, the collegiality and the brutal competition, academic tenure, grantsmanship, the Nobel Prize . . ." constitutes an incomplete list of topics that I cited and then covered in my first novel of a projected tetralogy, *Cantor's Dilemma* (New York: Penguin, 1991); topics that I claim in that book to be "the soul and baggage of contemporary science." Indeed, there are even more esoteric topics—all related to our standards and behavior—that might be presented under the guise of fiction. The next chapter gives a specific example.

22 Mentoring: A Cure for Science Bashing?

"Do you know the word Schadenfreude?"

"No."

"It's one of those German words, like Gestalt or Weltschmerz, that has a special flavor that doesn't come across in the English equivalent: gloating. Schadenfreude is more subtle and yet meaner. The more impeccable your reputation, and the more significant the work you retract, the greater the Schadenfreude among your peers."

"I can't believe what you're saying," exclaimed Paula. "You scientists, you upholders of the social contract, gloat like other mortals when somebody makes a mistake? Even when he confesses the mistake?"

Cantor let out a sigh. "I'm afraid the answer is yes. I've been guilty of that. I mean gloating," he added quickly. "I've never had to retract any of my published work and I hope to God I won't need to do it this time. Because of the rareness of such events, innocent or otherwise—"

"What does 'otherwise' mean?" interjected Paula.

"Data manipulation, even outright fraud. . . ."

"Does that happen?"

"Not often," he responded firmly. "But just as I said, precisely because of the rarity of such retractions, people's memory is unbelievably long: I'd guess a

Note: Original text written in 1991.

lifetime. Because of our mutual dependence and our need for absolute trust, once somebody's credibility in science is damaged, it can never be totally repaired. Most often, it's gone for good."

This excerpt comes from my novel, *Cantor's Dilemma* (Penguin, 1991). In it the fictional hero, Professor Isidore Cantor, a molecular biologist about to get a Nobel Prize for his theory of tumorigenesis, refuses to wear the hair shirt of a public retraction of a recently published, non-replicable, and possibly fraudulent experiment by his favorite postdoctorate fellow, Jeremiah Stafford.

Many readers have asked whether the two principal characters are modeled after David Baltimore—also a Nobel Laureate and molecular biologist—and his colleague, immunologist Thereza Imanishi-Kari, whose problems with the scientific community, U.S. Representative John D. Dingell's subcommittee, and the Secret Service made headlines. My answer is no. Except for one social gathering many years ago, I had never formally met Baltimore prior to writing this essay. His research and that of Imanishi-Kari are far removed from my professional competence.

Furthermore, the manuscript of *Cantor's Dilemma* was in my publisher's hands in 1988, and the short story on which the novel is based was published in 1986 in the *Hudson Review.* Yet Isidore Cantor could have spoken for all the David Baltimores (or Carl Djerassis, for that matter) when he responded to his nonscientist friend Paula Curry, whom I used as my archetype of the general public.

"What do you people expect from each other? Absolute perfection?" exclaimed Paula.

"Of course not. But if the work is important, if it influences the thoughts or research direction of many others, the accusation would be: 'why did you publish in such a hurry? Why didn't you wait until your results were validated?' "

"And what would your answer be, if you were asked? Why did you publish in such a hurry?"

"To be quite honest, most scientists suffer from some sort of dissociative personality: on one side, the rigorous believer in the experimental method, with its set of rules and its ultimate objective of advancing knowledge; on the other, the fallible human being with all the accompanying emotional foibles. I'm now talking about the foibles. We all know that in contemporary science the greatest occupational hazard is simultaneous discovery. If my theory is right, then I'm absolutely certain that, sooner or later—and in a highly competitive field like mine, it's likely to be sooner—somebody will have the same idea. A scientist's

drive, his self-esteem, are really based on a very simple desire: recognition by one's peers. That recognition is bestowed only for originality, which, quite crassly, means that you must be first. No wonder that the push for priority is enormous. And the only way we—including me—establish priority is to ask who published first. Suddenly you seem very pensive, Paula. Did I disappoint you?"

Of course, Paula Curry was disappointed. She and the general public would be even more disappointed if Cantor had also talked to her about the seemingly never-ending and ever more time-consuming involvement in grantsmanship, the ever more difficult chase for research support, which is another reason for the competitive drive mentioned by Cantor. Even some scientists disliked what I said through Cantor's words.

I do believe that academic research among the scientific elite for which Isidore Cantor is a prototype, especially in highly competitive fields such as molecular biology or immunology, has acquired aspects of a rat race. If Cantor's justification for the behavior of many academic superstars in cutting-edge disciplines is accepted, then Representative Dingell and other critics with their district attorney approach have wasted their time or, at least, the taxpayer's money. I say that because a policeman's approach in dealing with important problems of social deviance usually addresses only the crime already committed, hardly ever its underlying cause. This is especially so when the policeman or district attorney has only the vaguest notion about the culture and behavior of the group to be policed or punished, and when the penalty may be more costly to society than the perceived crime.

Science is both a disinterested pursuit of truth and a community, with its own customs, its own social contract. Members of that community pride themselves on their abstract languages and their tribal customs. Their tribal behavior is acquired largely by intellectual and cultural osmosis from their mentors and their peers, rather than from textbooks.

The ethics and conduct of research are hardly ever taught in formal courses. They are acquired in a mentor–disciple relationship that affects the very manner in which we speak and write about our work. Written scientific discourse, by its very nature, is monologic. I shall use the dialogic discourse of the novelist to illustrate one of our most common idiosyncrasies: coauthorship. This practice among scientists is frequently incomprehensible even to academics in the humanities, let alone to the lay public.

"I put my best young collaborator, a Dr. Stafford, on the project. . . . But I drove him very hard, I'll admit that. I was so convinced that the theory was right, I did something that under ordinary circumstances I would never have dreamed of doing. I basically told the man that he had to finish the work in three months."

"And did he?"

"He did. We published the work.—"

"We."

Cantor looked puzzled. "Yes, we. Why do you ask?"

"Well, if he did the work, why did you publish it with him?"

"God, Paula"—he sounded annoyed—"we do have a cultural gulf to bridge. I don't want to spend the time on it now. Let me just assure you that in science it's de rigueur. ***I*** *thought of the problem and the solution,* ***he*** *did the actual work, and* ***we*** *published it together. That's how it's always done."*

Mentor–disciple relationships, not uncommon in other endeavors, are particularly intense and potentially long-lasting in scientific research. Frequently, the umbilical cord between mentor and protégé is never cut, not even after the mentor's death. As in any intense relationship, the consequences of such intimacy can be both nourishing and pernicious. When the scientific elite displays occupational deviance, we pay a double penalty: The scientific enterprise is tainted and so is the example we set our disciples.

I have no intention of judging the highly publicized incidents in the laboratories of superstars such as David Baltimore or Robert Gallo—given their sub judice status and my own lack of detailed knowledge of the circumstances. Nor do I wish to rehash some of the highly publicized instances of scientific fraud such as the John Darsee or Mark Spector cases of the 1980s. But all of them display some common features. The laboratory chiefs are world-renowned scientists who, prior to those incidents, made enormous intellectual contributions that have withstood the test of time. The research in question was performed in elite institutions on some of the hottest topics of contemporary science. And finally, the research groups for which these senior scientists bore the financial, administrative, and entrepreneurial responsibility ranged from large to enormous. It stands to reason that the size of a research group is inversely proportional to the time available for a meaningful direct mentor–disciple relationship.

Certainly ambition, Nobel lust, and/or competitiveness—precisely the factors that fuel great scientific successes—were the infectious causes. In the final analysis, however, most of the incidents that have generated so much publicity can be reduced to a breakdown of the mentoring process. For it to be improved the community itself, and not a Congressional policeman surrogate, needs to face the issue. As I found out in my own discipline of chemistry, this is easier said than done.

For most of my research career, I have led concurrent academic and industrial lives, both of them spanning decades. Over that period I have interacted in each setting cumulatively with several hundred collaborators, thus providing the experiential background upon which I base my personal conclusions.

Although industrial research suffers from many problems, these are on the whole quite distinct from those incidents in academia falling under the rubric of "hoaxing, forging, trimming, and cooking," as defined in 1830 by Charles Babbage in his famous admonition to the Royal Society. Industrial researchers are not necessarily more perfect than their academic counterparts. But deviancy in their occupational behavior generally manifests itself along different lines, because much of the recognition and reward structure in industry is provided within the internal environment through promotions, perks, and financial remuneration. The temptation or, indeed, opportunity for premature or concocted publication is rare. Academics, on the other hand, by definition display their virtuosity as well as their excesses and aberrances in public, because the award they crave most is external peer recognition.

I have spent more than four decades in academia as a mentor of graduate students and postdoctorate fellows, the latter group coming from more than 50 countries. Right from the beginning, in 1952 as an associate professor at Wayne State University, I started with 14 collaborators. And for at least three decades, my research group consisted of 20 to 22 individuals.

Thinking back to that time, when I daily dealt directly with each student or postdoctorate fellow, I seem never to have questioned the quality of my performance as a mentor. I never consciously examined that key role of mine, nor did I have a formal yardstick by which to measure that performance.

In 1987, as a member of the National Academy of Sciences' Institute of Medicine's Committee for the Study on the Responsible Conduct of Research, I chaired the panel on education and training for research. Its rapporteur, Jules Hallum, then chairman of microbiology and immunology at Oregon Health Sciences University, subsequently became director of the National Institutes of Health's Office of Scientific Integrity and thus in charge of the investigations dealing with the Baltimore and Gallo cases. Other panels covered issues ranging from clinical research standards and practices to authorship, referee, and publication practices.

I was the only chemist on the 17-member committee, which consisted predominantly of biomedical and clinical professionals. The committee's formal mandate was to "assist NIH, other government agencies, professional societies and journals, and universities in formulating policies and procedures to improve the integrity and quality of biomedical research."

One theme, brought up by every panel and surfacing as the central issue in our education and training for research panel, was the quality of the mentor–trainee relationship. Particularly noted was its apparent deterioration in light of larger research groups, greater competitiveness, and more time-consuming administrative and financial responsibilities of senior scientists at major U.S. research universities.

During the 12 months of our deliberations, I could not help but reflect on my past performance as a mentor. I especially focused on how it might be evaluated in light of today's problems, many of which were unheard of in the 1950s and 1960s, unheeded in the 1970s, and still ignored in the 1980s. I was particularly struck by the ad hoc manner in which many senior professors (including me) in the top chemistry research departments deal with the mentoring issue: Young faculty members get absolutely no formal guidance, it being assumed that their own ad hoc graduate school and postdoctoral experience will turn them into skilled, wise mentors. More important, I was struck by the total absence (at least in those elite institutions with which I am familiar) of any formal mechanism for evaluating the mentor's performance.

In many respects, the mentor plays the role of intellectual parent in the development of a new researcher. Few parents ask for feedback from their children; or when they get it, most attribute it to rebellion or lack of respect. But should this be the attitude of intellectual parents who meet their intellectual progeny only when the latter has entered adulthood?

Many chemists, when faced with the actual or implied ethical misconduct cases that have received so much notoriety, brush these aside with the comment that they are limited to biomedical disciplines. Is this because our field is not so "hot," our funding less competitive, our research more rigorous, our mentoring more effective?

As far as mentoring is concerned, how many chemistry professors—especially those with large research groups—have ever solicited anonymous feedback from their graduate students? In this age of virtually universal, anonymous student evaluations of a professor's teaching ability and performance, why is no assessment done for the most important graduate teaching aspect of an academic—the imprinting of the disciple's research persona?

As an initial experiment, I constructed a questionnaire with the preamble, "When you joined your professor's research group, did you discuss with him or her the following topics?"

1. The nature of the mentor–trainee relationship
2. The publication policy; that is, the order of authors, who else might be included as coauthor, who would write the first draft, and so forth
3. Record-keeping procedures, such as type of notebook used, retention of samples, and spectra
4. Any ethical questions concerning the conduct of research
5. The university's patent policy
6. The professor's views on patenting possible discoveries, and your role in such patent applications
7. Ownership of lab notebooks and other primary data, such as spectra and samples
8. Consideration of formal statements on the ethical conduct of research prepared by Sigma Xi or the National Academy of Sciences
9. Possible courses or more formal discussions of such issues

The nine questions could be answered in a couple of minutes by simple yes or no answers. In addition, after each question I added the phrase, "If your answer is affirmative, was this discussion conducted formally (answer yes or no) or informally?"

Distributing such a questionnaire among the top ten U.S. chemistry departments would provide hundreds of data points on which to draw some conclusions, the most important of them being whether more formal lecture or teaching mechanisms for the proper conduct of research are desired. My proposal to faculty members of one of our most prestigious chemistry departments to distribute such a questionnaire was met by surprising responses: outspoken opposition or total silence. The chairman's summary of the faculty's attitude was pithy: "The proposal may exacerbate rather than ameliorate some tensions that always exist within research groups." I do not believe for a moment that this reply was sui generis. Rather, I suspect that a similar response would have come from the majority of the stars in the chemical research firmament of U.S. universities.

Independent of the distribution of such a questionnaire, I also would propose—this time to the cognizant deans to whom science research departments report—implementation of an annual anonymous evaluation system. This assessment would ask questions similar to those in standard teaching evaluations, only in this instance encompassing the various components of responsible mentoring.

It would be naive to suggest that such steps will immediately or even in the near future resolve the most vexing problems of ethical misjudgment, impropriety, or misconduct. Over the long term, however, such attention to the mentoring process cannot but help upgrade ethical standards. Careful, confidential, and compassionate analyses of such evaluations may help to pinpoint potential problem areas. Such analyses also may help to answer one of the most sensitive questions in a research university: When is a research group too large for adequate quality mentoring without introducing a layered system of "lieutenants" with concomitant loss of collegiality and significant personal interaction?

Most of the recent ethical problems that explicitly or implicitly have been raised by the media, by various government sources, or by the public at large, have been associated with the crème de la crème of the academic community. Scientific research is never a populist activity. Difficult and superb research is invariably performed by the elite, usually at elite institutions. Elitism invariably engenders jealousy. It is no wonder that *Schadenfreude* flourishes when the elite stumbles, but at least there is a touch of justice in such gloating when it focuses on the misdeed of a privileged individual: the principal scientist. The high moral tone of such gloating is related to the perception that any laboratory digression is ultimately the principal investigator's responsibility, because the appellation "principal" carries with it both privilege and duty.

Much of the public controversy surrounding the Baltimore affair, though focusing on his role as principal investigator, is infected by acute *Schadenfreude.* Is, then, Baltimore bashing also science bashing?

In my opinion, the *Schadenfreude* component of Baltimore bashing is bound to contribute to science bashing. So will the virtually exclusive attention to the police aspects of the case and to the desire for an unabashed mea culpa by Baltimore, the most famous of the five authors (Weaver, Reis, Albanese, Constantini, Baltimore, and Imanishi-Kari) of the 1986 paper in *Cell.* Absent from discussion is a debate over the underlying causes of this incident. I do not object to a fair and thorough investigation of the scientific aspects of the *Cell* paper. Nor do I question a critical examination of the presumed cavalier treatment directed at the whistle-blowing postdoctorate fellow, Margaret O'Toole.

Rather, I am bothered by the insufficient attention given to fundamental preventative measures such as improved mentoring that could deal with such behavioral deviance. Even Paul Doty, noted Harvard professor emeritus of biochemistry, in his critique [*Nature,* Volume 352, page 183 (1991)] of the "Weaver et al." case (a sanitized synonym for "Baltimore case") ended with a clarion call to the scientific community to reassert its ability to police itself, without, however, offering a cure. Doty's commentary elicited from Baltimore an open "Dear Paul" letter [*Nature,* Volume 353, page 9 (1991)] focusing primarily on some debatable experimental issues [since adjudicated by independent confirmation in Baltimore's favor] and containing the following sentence: "You say that I failed traditional standards of science but you have not discussed the events with me, choosing to rely on information from others."

Why did two distinguished mentor role models—on a first-name basis with each other and, until recently, geographically separated by less than one mile—have to conduct such a dialogue in the pages of *Nature* rather than in person? As the Doty–Baltimore dialogue highlights, trust and collegiality—two vital attributes of a successful mentor—receive short shrift. If we do not center our attention on the importance of these qualities and on the mentoring process itself, science bashing will not only become more pervasive, it may even be deserved.

23

Basic Research: The Gray Zone of Professional Bigamy

In these days of societal concern about practical applications of scientific research, the editorial columns of *Science, Nature*, and other journals argue the need for continued generous support of basic research in academia. We can be certain that some basic research will have practical value, but we cannot confidently predict which of these advances and areas is destined to have the most impact. Some borders have been drawn by stating (editorially) "that in the administration of basic research, the ultimate question is strategic priorities." But what about the gray zone—when a basic research discovery moves toward practical implementation?

Explicitly or implicitly, the assumption is always made that academic eggheads in white coats should leave that job to industrial entrepreneurs. But dramatic changes are under way, and nowhere more strikingly than in the field of biomedical applications. Many newspaper columns have noted the pervasive connection of most top biomedical researchers in American universities to industry, but they invariably criticize the associated financial rewards, the implication being that money in academia unfailingly corrupts. We tout America's thriving entrepreneurship (almost entirely based on prospects of financial gains) while deprecating flourishing academic entrepreneurship, forgetting that the biotechnology industry in America would never have taken off

Note: Original text written in 1993.

without the active involvement of academic investigators in hundreds of fledgling enterprises. Associated rewards in terms of stock options or stock ownership (standard currency in any industrial entrepreneurial setting) invite instant suspicion and criticism; the position of the Howard Hughes Medical Institute that Stanford's Irving Weissman resign his position after he founded Systemix Inc. is a dramatic example.

Let us not pretend that potential conflicts of interest and even egregious examples of academic misconduct are caused primarily by yearning for financial gains. Nobel lust or the craving for a multitude of lesser kudos are most commonly responsible for academic deviance, but I have yet to learn of an academic code of conduct addressing impermissible levels of personal ambition. The Hughes Institute would hardly have objected if Weissman had launched his discovery solely by way of the balloon of a scientific publication, even if the latter had then led to a Nobel check exceeding $1 million. The Hughes Institute would hardly have blinked if some industrial enterprise, say in Japan or Europe, had used these published results, uncontaminated and hence unprotected by a patent application, to develop human cancer therapeutics and reaped all ensuing financial benefits.

Why object automatically if the academic discoverer wishes to continue shepherding his or her scientific baby along the road to practical maturity, prompted in part by financial gains? Why should such person have to abandon the academic laboratory to do so? Monogamy is great for stable marriages, but what is the evidence against the benefits of intellectual bigamy in academia (with its associated financial benefits to the individual and eventually to society)? More than half of our graduate students and postdoctoral fellows pursue careers in industry. Could a professor with active participation in the extremely complicated, multidisciplinary approach to practical realization of laboratory discoveries not be a better mentor? Could an academic, serving in some part-time directorial or managerial position in industry, not offer a perspective rare in conventional businesses?

These are not theoretical questions, as my own experience attests. Years before the biotechnology explosion, I straddled both sides of a then much less penetrable wall by serving simultaneously as a chemistry professor at Stanford University and as an officer (including chief executive officer) of research-intensive industrial enterprises; the industrial position carried handsome compensation in terms of salary and stock options. There are other examples of such professional bigamy (most commonly disclosed in annual proxy statements) that have resulted in direct benefits to a wide community. I estimate that my own industrial activities, during my concurrent academic service, were responsible for creating several thousand jobs, most of them highly technical, in the San Francisco Bay area. None of this, of course, changes the perception (often by sanctimonious critics) of the corrupting influence of money received in return for intellectual services—a truism applicable to so many facets of contemporary society. The only feasible safeguard for society is open disclosure. Open professional bigamy, with all the associated legal responsibility, is far preferable to hidden affairs disguised under ambiguous terms such as "consultant" or "adviser."

I conclude that encouraging, rather than condemning, professional bigamy among academics can be societally beneficial, provided it is accompanied by clearly defined guidelines. Such guidelines should cover the following topics:

1. *Time limits for outside activities by full-time academics.* The limit commonly imposed by universities is the one-day-per-week equivalent. It is unrealistic to enlarge this, nor is it generally feasible for an academic to assume outside managerial responsibilities under such a time constraint. The latter constraint appropriately restricts one's industrial activities to memberships on boards or to conventional short-term consultantships.

2. *Nature of permitted outside activities.* Rather than indulging in the maxim, "what is not permitted is proscribed," I would list the few activities that are always off limits and then set up an institutional mechanism that would handle on an ad hoc basis all other questionable practices. Examples of invariably prohibited activities include the utilization of university property, facilities, or personnel (notably students) for the benefit of the company with which the academic is associated; evaluation of actual or potential products for eventual government approval (a notable example being phase II and especially phase III clinical trials of experimental drugs) or promotional purposes; and the restriction of free publication of university-conducted research.

3. *Length of unpaid academic leave.* Most universities have pertinent rules, ranging from the most common two-year period (for example, Harvard University) to open-ended arrangements, the latter usually associated with important position in government (as if power were not as corruptible as money). I would favor specified limits for such leaves, irrespective of the justification.

4. *Scope and nature of part-time positions.* Most of the institutional precedents pertain to medical schools with their part-time clinical appointments, which are justified on professional grounds: to find teachers who can tell students about the real (clinical) world. This argument should be invoked for most other professional areas in order to encourage, rather than restrict, part-time academic positions once the quality control of approved tenure has been passed. The academic commitment should not go much below 50% in order to ensure adequate physical presence on campus; such a level would still permit meaningful involvement even at the managerial level in many technical enterprises. Outside managerial or directorial responsibility could be restricted to areas having a direct bearing on the academic's professional field of competence, as is the case with all clinical appointments.

Entrepreneurship could flourish under such liberal, part-time rules with associated societal benefits. Nor is there any evidence that such entrepreneurial activities by responsible part-time academics impair in any sense the quality of their teaching or research. In these days of constrained university budgets, increased numbers of part-time, tenured faculty may help in many ways. For instance, because many universities pay no fringe benefits below the 50% level, a 49% level of academic commitment would relieve the university of a substantial financial burden that is usually assumed by the outside employer. Because such part-time faculty is more likely to come from scientific or engineering disciplines, the savings could be funneled to the arts and humanities, which are suffering in the current economic climate.

Let us not overlook how many studio art and creative writing programs are proud to claim well-known artists, writers, or poets as part-time university faculty. In contrast to the case for scientific collaborations, hardly anyone has raised the question whether universities are entitled to a percentage of the sale of high-priced items of art or to copyright and a share of royalties of best sellers from such faculty members.

5. *Patent policy.* The institution's patent and royalty policy is the origin of most conflicts of interest and of potentially the largest monetary rewards. There is a viable alternative for the range of restrictive or free-wheeling patent policies currently found among American universities: A Stanford University committee, under the chairmanship of biologist Craig Heller, has drafted a more precise conflict-of-interest policy, which has received considerable media scrutiny. One proposal encompasses the requirement that all inventions made by Stanford faculty become

automatically the intellectual property of the university unless it decides not to seek patent rights. This is precisely the situation under which virtually all full-time research personnel now operate in industry. Assignment of patent rights does not carry royalty benefits for such industrial inventors, presumably because practical invention is a key component of industrial employment. Because that is not the case in academia, royalty payments from any university-owned patents to the inventors are appropriate, but with the university retaining a prospectively defined percentage. At present, many universities do this for those patent applications that a faculty member chooses to file. Under the recommended scheme, this choice will be made unilaterally by the university. The possibility of licensing such patents exclusively to some corporation already exists in many university patent policies and thus would not inhibit inventor–faculty involvement with new or existing companies.

Objection to such exclusive licenses is often raised on populist grounds (why should the taxpayer's money benefit one company?) without realizing that in many technical areas, a company would not enter into royalty-bearing licenses of some basic research discovery without such exclusivity in view of the risk and extraordinary cost associated with bringing such discoveries to practical fruition. In the absence of such exclusive licenses, the taxpayer's original investment in basic research would benefit no one. A substantial element of conflict of interest would be removed if all such decisions were made at the university level with appropriate disclosure.

6. *Resolution of conflicts of interest.* It would be naive to assume that this list of recommended guidelines, or indeed a much longer one, would prevent all conflicts of interest. One helpful step would be to establish an office of a special ombudsperson, with experience in the academic and industrial world, to whose attention potential conflicts of interest could be brought in confidence at an early stage by any party.

I can think of few better ways to stimulate societal responsibility for one's basic research than to be formally involved in the necessary technology transfer from the laboratory to the ultimate consumer. The perception of the corrupting influence of money cannot be changed, but because it applies to virtually all areas of contemporary society, why not focus primarily on reality rather than perception?

24 Some Forms of Art Patronage

The word "some" is the operative term of my title. Just because I shall emphasize certain forms of art patronage, however, does not mean that I denigrate others, notably those extended by government and corporate sources. But given current controversies about the dispersal of what is crudely called the "taxpayer's dollar"—with all the connotations of populist censorship and parsimony inherent in those words—I shall limit myself to the most nonbureaucratic version: private patronage.

All art appreciation, be it of "The Arts" or of individual artists, is by definition subjective. The role of personal taste is particularly pronounced in three common forms of private patronage of the arts: acquiring, collecting, and commissioning. It is not surprising, therefore, that the most avant-gardist and daring artists are likely to be least patronized, primarily because their works issue from an exploration of unfamiliar aesthetic or intellectual territory.

COLLECTING AND COMMISSIONING ART

All collectors acquire art, but not all acquirers of art are collectors. Usually, collecting requires components of knowledge and aesthetic judgment that are often absent in the simple acquisition of a work. Col-

Note: Original text written in 1991.

lectors are patrons of artists if the latter are still living. But when the artist of the collected work is dead, and especially when the artist has been fetishized, we are entering the territory of the Arts with all the connotations of canonization that the capital letter A implies. My experience with collecting the works of Paul Klee is such an example, and I cite it because promising my Klee collection to the San Francisco Museum of Modern Art has made me into a kind of patron of the Arts. Concern with patronage of the artists came only later.

My own definition of art collecting is somewhat narrow. I am not referring to the purchase of occasional pieces of art, nor to collecting based primarily on investment motives. Rather, I focus on the serious collector, the person who concentrates on a specific artist, a specific art movement, or applies some other self-imposed criterion; who renders an intellectual judgment and to that extent places a personal signature on the collection. Assembling five Picassos at random is very different from deliberately selecting five Picassos to make a specific aesthetic, pedagogic, historic, or personal point about the artist.

Collecting the works of dead artists becomes a form of patronage only when it serves the public benefit. In many respects, the serious collector of a dead artist's work also becomes that artist's interpreter. If one takes this role seriously, then such collecting can become an exciting creative process: One presents a special view of the artist by selecting specific aspects of that artist's output. Displaying many works of one artist in one place is, in my opinion, far preferable to distributing the works over many sites. When such a collection is made available to the public, the social benefit is clear.

I have a long-held personal belief about private art collections—especially those specializing in one artist. If a significant portion of an artist's output is concentrated in one collection, then it ought to be available to the public. A museum is the obvious place, but all too often this means the basement of the museum rather than exhibition space. Because of Klee's modest space requirements—he is the perfect example of a master of the "petit format"—one need not unduly impose on a museum's limited gallery space by requesting that a significant portion of the collection always be on display. In fact, the size of his works requires intimate space, and intimate space immediately leads to intimacy of another nature: close inspection of the work, attention to detail, and the ultimate punctuation mark, the gasp of delight.

Many cities of San Francisco's size, both in this country and abroad, are frequently villages—in a cultural sense. The cosmopolitan character of San Francisco, its geographical location, and its history have helped to avoid giving it the cultural village status. It is a city on an intimate scale, which deserves an intimate painter like Paul Klee, and this is where my collection will remain.

One of the pleasures of collecting art is that it need not be justified to anybody else; thus, I shall comment only briefly on how I became enamored of Klee, whose works I have collected for almost 30 years. In my college days, I had seen reproductions—on postcards, calendars (modern society's ultimate accolade), posters, catalogs, art books—of Paul Klee's most famous works, such as "Twittering Machine" or "Ad Parnassum." Subsequently, I saw many of his oils, drawings, and watercolors in museums in Europe as well as in this country. Only later did I learn something about the man through his notebooks, diaries, poetry, and music. That subsequent knowledge only confirmed my intuitive response gained from his art: Here was an extraordinary combination of intellect, aesthetic sensitivity, and desire for experimentation. Over the course of four decades, Klee created nearly 9000 works of art in almost every medium—from graphics to oils, from surfaces such as burlap, plaster of Paris, and glass to every type of paper and canvas, and even puppets and sculpture. In terms of subject matter and technique, he seems to have anticipated almost every art movement, even those of today.

In the mid-1960s I went to my first Klee show in a commercial gallery, where all of the works were for sale. Until then, actually owning a Klee had not occurred to me. But included in that show in London were two magnificent watercolors from his *Bauhaus* years in the 1920s—rather large ones for an artist who usually worked on such a small scale—to which I kept returning over and over again. "Should I? Could I?" I asked myself. Finally, I approached one of the gallery employees and asked about the price. "The 1925 'Horse and Man'?" he asked, looking me up and down. "Sixteen," he finally said.

"Sixteen what?" I wanted to ask, but didn't. I knew it could not be 1600, and was unlikely to be 160,000, so it had to be 16,000. But 16,000 what? Dollars, pounds, or even guineas? "And the other one, the 1927 '*Heldenmutter*'?" I asked hesitantly.

"Eighteen."

"Hm," I replied and went back to look at the pictures. A few minutes later, the man appeared by my side. "Which one do you prefer?" he asked in a slightly warmer tone.

"I can't make up my mind," I said, "both of them are superb."

"Buy them both," he said matter-of-factly, "and maybe we can arrange a better price."

Bargaining, whether in a marketplace in Mexico or a bazaar in Cairo, always makes me uncomfortable, but this time I haggled by default. Every retreat of mine, every inspection and reinspection of first one, and then the other Klee, caused the price to drop. They were not big reductions, but given the overall sums—far above anything I had ever spent before on art—they were not insignificant. Finally, I said, "I'll have to think about it." A couple of days later, I was the owner of not one but two Klees. By now I own more than 100 of his works in various media, including most of his early graphics, but these two watercolors are still among the *crème de la crème.*

A few years later, I visited Felix Klee, the painter's son, in his flat in Bern and saw his extraordinary collection, which included puppets his father had made for him—a genre of Klee that until then was unknown to me. He also showed me his mother's guest book. The first entry was Wassily Kandinsky, who did not just inscribe the book, but drew a colored picture on that page. Not to be outdone, many of the other guests—Lyonel Feininger, George Grosz, and others I do not remember—did likewise. It is one of the most intimate and exquisite documents of European art of the 1920s and 1930s.

The type of patronage that I believe is needed most urgently at the end of the twentieth century is direct support of artists, notably by commissioning works of art. The Medicis are the type of patrons I admire most, because they supported a multiplicity of artists as well as disciplines: painters, sculptors, architects, humanists. Giorgio Vasari commented in 1550: "Cosimo de'Medici was the perfect patron. He kept Donatello continually at work, and Donatello understood Cosimo so well that he always did exactly as Cosimo desired." Although the aesthetic preferences of the patron may unconsciously or deliberately influence the artist, I believe that most artists would be more than willing to take that risk in return for long-term, enlightened patronage.

Another Italian living in Santomato di Pistoia, only a few miles from Cosimo's Florence, exemplifies the impact that just one private art patron can exert. Some years ago, Giuliano Gori—a textile manufacturer in Prato, next door to Florence—acquired the estate and olive groves of the seventeenth-century Cardinal Fabrone in the Tuscan hills above Santomato. Named *Fattoria di Celle,* this has become in less than a dozen years one of the most ambitious private collections of contemporary art in Europe. Gori's patronage encompasses collecting as well as commissioning, but it is the latter activity, in which he indulges on a grand scale, that I wish to emphasize.

With a handful of exceptions, all of the outdoor installations, spread over many acres of deciduous forest, olive groves, and vineyards, have been commissioned. There is no common denominator of medium, size, national origin, or artistic school. The artist is invited to create a site-specific work of art with much of the labor provided by workers from the premises or the neighborhood. Describing these works is largely pointless, because words cannot do justice to the visual impact that inherently "nonenvironmental" materials (such as concrete or Corten steel) can have in such a natural setting when utilized sensitively by a creative artist. I know of no public museum that has gathered so much *commissioned* contemporary sculpture as the private *Fattoria di Celle.* The latter does not suffer from the financial restrictions of most museums; furthermore, aesthetic decisions are made by one person and not by committees, which often sacrifice the very best on the altar of consensus.

Consider the range encompassed by the following artists, my selection of the 29 sculptors whose works grace an area of approximately 70 acres within the immediate vicinity of the main Villa Celle and the adjacent *Fattoria* building (itself a monument to indoor commissioned art): Dani Karavan's (Israeli) white concrete line, starting in an open field, seemingly passing through two plane trees and then penetrating a thick, planted bamboo grove (hardly expected in Tuscany) down to a small lake, where a row boat—part of the Karavan installation—is moored; Robert Morris's (American) green and white marble labyrinth, set at an incline, thus introducing a startlingly effective spatial component to the usual puzzle of a labyrinth; the multicomponent marble "Death of Ephialtes" by Anne and Patrick Poirier (French) at the foot of a small waterfall centered around a huge eye penetrated by an arrow; Magdalena Abakanowicz's (Polish) deeply moving assembly "Catharsis" of 33 bronze hollow torsos in a fenced field facing a stone wall; and Alan Sonfist's (American) "Circles of Time," a huge circle of local rocks surrounding a circle of laurel, followed by a circle of bronze castings of endangered and extinct trees, and finally ending in a small planting of olive trees and wheat. Close by is Inoue Bukichi's (Japanese) "My Sky Hole"—a stone path descending into the earth through a glass door into a concrete tunnel, illuminated by small sky lights and eventually rising into a glass cube through which the visitor steps out into the open—a work that took more than four years to complete.

Just as I started my selection with work made out of concrete (Karavan's straight white line), I end with another installation from such industrial material, Mauro Staccioli's (Italian) enormous triangular slab, pushing through the bush toward the passerby. Not too far away, within easy sight, is a charming 19th-century teahouse. The meadow below it forms a natural amphitheater, which has been converted into an outdoor performance space and stage by the American sculptor Beverly Pepper, who resides in Italy.

Starting in 1991 multidisciplinary performances were staged in this space, which is testimony to Gori's wider-ranging patronage. Indications of such interest can be discerned in the extraordinary indoor space, which Gori has created during the last decade; first in the main villa, where he resides, and more recently in the adjacent *Fattoria,* which consists simply of numerous white-washed rooms and open spaces, illuminated by super-modern Italian mercury lights. Each room, corridor, or entrance way has been offered to one of 34 artists (the majority—such as Mimmo Paladino, Michelangelo Pistoletto, and Aldo Spoldi—Italian) who have created an array of installations, ranging from minimal and whimsical pieces to extraordinarily complex and moving ones. Luciano Ori's visual concert "Pictures at an Exhibition: Concert for a Year" constitutes a visual artist's counterpoint to Mussorgsky's musical "Pictures at an Exhibition." The center of the room contains a grand piano and a life-size cutout of a musician—a silk screen on glass so realistic as to fool most spectators. The accompanying pictography covering the walls contains 15 works of art on music sheets created by a "painter," who changed into a "visual poet" and finally into a "technological painter." This concert cannot be performed, yet it is being performed in front of the viewer.

SPONSORING THE ARTIST

Even the most enlightened commissioned art passes through the filter of the patron's taste. Furthermore, few patrons commission works across the spectrum of the arts to embody also literature, music, and choreography. The simplest solution to that problem is to have more Goris of music, Goris of poetry, Goris of. . . . But there is another form of direct support of artists that can encompass the entire range of disciplines under one roof and do so with a minimal impact of the patron's personal taste upon the artist's style or output. In the final analysis, the sole criterion for sponsorship should be confidence in the quality and integrity of the artist, which can be accomplished by providing the most precious gift: undisturbed time to pursue one's art, with no strings attached to the artistic output. The resident artists' colony, a peculiarly American form of contemporary arts patronage, fits that definition.

Yaddo in upstate New York, MacDowell in New Hampshire, and the Rockefeller Foundation's Villa Serbelloni in Italy are three well-known examples that have existed for decades. I shall concentrate on a more personal example, the Djerassi Resident Artists Program near San Francisco, which has, in a little more than a decade become one of the largest resident artists' program west of the Mississippi, by now having supported close to 700 artists. The most interdisciplinary of all such American programs, it offers (for periods of one to three months) free room and board as well as specialized studio space to 10–11 artists at a time in the disciplines of choreography, dance and performance art; literature, such as fiction, poetry, drama, biography, and criticism; music, with major emphasis on composition, but not excluding performance; and the entire range of visual arts encompassing sculpture, painting, drawing, film, photography, video, ceramics, and fiber art.

The program's property of more than 600 acres (carved out of my family's 1200-acre SMIP Ranch) overlooks the Pacific Ocean; it is about an hour's drive south of San Francisco in a spectacular setting of rolling range land and redwood forest in the Santa Cruz mountains. This site offers, like Giuliano Gori's Celle, a major opportunity to sculptors interested in large-scale, site-specific projects. I chose this example of private patronage because I am most familiar with it: The program was initiated as my response to the suicide of my daughter, who at 28 was a promising visual artist and poet. It took my daughter's suicide to make me take seriously the patronage of the living.

A record of the founding of the Djerassi Resident Artists Program has appeared in my autobiography in a chapter entitled "A Scattering of Ashes." The program's quantitative achievements are summarized in Table 1, but such statistics are meaningless without a sense of the quality and variety of the work that can arise from such patronage. The artistic examples cited are not based on my personal preference, but rather were picked because the works illustrate the interdisciplinary cross-fertilization that is possible when artists from many disciplines, countries, and cultures are brought together for a few weeks or months.

Applications for residencies are solicited from artists at two levels. One is the level of great promise: artists who have a record of solid achievement but are not yet well known, and for whom an appointment as a resident artist might contribute to professional advancement.

Table 1. Djerassi Resident Artists Program Statistics

Category	1979–82	1983	1984	1985	1986	1987	1988	1989	1990	1991	1992	1993	Total
Total residencies	14	52	53	46	50	70	67	92	75	53	68	45	685
Women	12	28	30	22	20	31	37	42	43	33	45	25	368
Men	2	24	23	24	30	39	30	50	32	20	23	20	317
Number of U.S. states	9	13	11	12	11	9	9	11	14	7	5	8	
Foreign countries	1	2	7	7	6	10	6	9	6	3	1	8	
Writers	5	26	27	21	24	23	28	31	28	19	11	16	259
Painters, drawers, printmakers	4	7	11	8	11	15	10	13	9	11	2	2	103
Sculptors, ceramicists	4	12	7	10	5	10	9	12	6	3	3	4	85
Photographers, media artists	1	1	2	0	0	2	4	8	5	1	4	5	33
Composers, musicians	0	3	5	4	7	11	12	9	8	4	2	5	33
Choreographers, performers	0	3	1	3	3	9	4	19	19	15	46	13	135

The other is the level of national or international distinction: artists with well-established reputations, for whom a change of scene in a collegial setting might offer intellectual refreshment and artistic inspiration.

With one exception, there is no requirement for any artist to leave behind any portion of completed work at the end of the residency. The exception was prompted by Klee's guest book, which his son Felix showed to me in Bern. Every artist, irrespective of discipline, is presented with a single, large sheet of drawing paper and requested to leave behind some artistic or intellectual statement. The collection of hundreds of such sheets constitutes the most personal of the many forms of archival documentation (photographs, videos, musical tapes, and tapes of literary readings) that the program has assembled. A single example from that "guest book," the last stanza from a poem by Amy Clampitt (1984), will suffice.

> COMING UPON AN UNIDENTIFIED WORK OF OUTDOOR ART
> NEAR WOODSIDE, CALIFORNIA
>
> the sudden, gladed glimpse of something made.
> Whose doing are these stationings
> of jute and brushwood, these totem-
> tripod harp shapes, that blood-red lyre whose
> silence stuns, among the redwoods—a holy
> place, an offering, a condition harking back to states of soul
> (an Eden's trespass slyly trailing withes of poison oak)
> one gropes to ascertain the name for?

The work mentioned in Clampitt's poem is a sisal and madrone construction by the British artist Patricia Leighton, a disciple of Magdalena Abakanowicz, whom this Polish artist (represented in the Gori Collection) had recommended to us in 1983. Entitled "Wake," the magic setting of these three constructions in a redwood clearing has not only influenced a poet. When the New York choreographer Rhonda Martyn was in residence in 1985, she choreographed—as part of a series of dances associated with outdoor sculptures—a performance around "Wake," which was presented to an invited audience on summer solstice (1985) at dusk.

Another impetus to a collaborative effort between choreographer and sculptor was provided by the Australian sculptor John Davis (1986), who uses eucalyptus branches as the armature, and painted canvas as the skin, of his large, anthropomorphic sculptures. His residency overlapped with those of the choreographer–dancer pair Clare Whistler (British) and Duncan MacFarland (American), who were so taken by the lightness and magic of Davis's sculptures that eventually, with the further collaboration of the Californian composer David Rosenboom, they completed a sculpture–dance project, "Systems of Judgment," which was subsequently performed (under the auspices of the Australian Arts Council) throughout Australia, at the 1988 Spoleto Festival, as well as in San Francisco.

At times, such collaborations are not hatched at the site, but planned ahead of time, as was the case in 1985 with a group of four artists from Basel, Switzerland. The choreographer Ester Sutter, the composer Joel Vandroogenbroeck, the painter–sculptor Andreas Straub, and the writer Aurel Schmidt—as part of their interdisciplinary project "Homenaje al Indio"—initiated and completed at the ranch a performance piece entitled "Energy Fields," consisting of five ritual sequences of dance for the environment. The audience had to follow the dancer along her outdoor path. Vandroogenbroeck's music, scored for piano, rhythm machine, and computer, included natural sounds (e.g., coyotes and crickets) recorded at the site. The performance was videotaped for subsequent showings in Europe.

To my knowledge, none of the other American artists's colonies includes choreography. The origin of our own dance program merits emphasis. The very first composer in residence, John Adams, spent three months in 1983 at the Foundation composing the synthesizer music ("Light over Water") for "Available Light"—the inaugural piece commissioned by the Los Angeles Museum of Contemporary Art from John Adams, the New York choreographer Lucinda Childs, and the Los Angeles architect Frank O. Gehry. Adams's comments about the difficulties of a transcontinental collaboration between composer and choreographer, and our own desire to promote multidisciplinary artistic endeavors, prompted us to apply to the James Irvine Foundation and the William and Flora Hewlett Foundation for funds to construct a dance and performance studio within the confines of our large, 12-sided studio structure. The generosity of these foundations, notably the additional funds provided by the Hewlett Foundation, led to the completion of several other specialized studio spaces: ceramic, photography, and music—each equipped with special equipment (e.g., kiln, enlarger, and computers) required by practitioners of these art forms.

The choreography space—by far the largest studio—has also served as the site for performances suitable for up to 100 spectators. Everything from premier dance, theatre, or music pieces has been presented over the years to supporters and visitors of the program. For instance, John Adams, who subsequently became a trustee of our program, presented his opera "Nixon in China" in video and lecture form in that space. Another occasion was a concert by John Adams, Ollie Wilson, and Ingram Marshall in front of specially designed, painted-canvas room partitions made by another artist in residence, Sandro Martini. At that event, these three former composers-in-residence played pieces composed by them at that site. Wilson (a professor of music at the University of California in Berkeley) had used part of a Guggenheim fellowship during his tenure at our program on a commissioned piece for the Boston Symphony, which was subsequently performed by Seiji Ozawa in Boston. Marshall, at that time (1984) a San Francisco composer, presented "Alcatraz"—a piece for piano, synthesizer, and slides of the abandoned Alcatraz prison.

Sandro Martini, the first of more than half a dozen visiting artists from Italy, overlapped during his residency with the American painter James Rosen (now the William S. Morris Professor of Art at Augusta College), whose métier at that time was the creation of contemporary, veiled versions of pre-Renaissance Italian paintings, such as Duccio's and Coppo's madonnas. Intrigued by his work, Martini proposed Rosen's name to the Museum of Contemporary Art in Ferrara, whose director invited Rosen as visiting artist for a few months. That stay was eventually extended to two years and resulted in a veritable explosion by Rosen of painterly homages to Italian Renaissance works, which were featured in a series of subsequent exhibitions—first in Italy and then in America. Rosen's variations, in hundreds of drawings, watercolors, and oils, of the Schifanoia frescoes in Ferrara, were one notable outcome of his Italian artistic sojourn, which was spawned in California.

As is to be expected, the overpowering landscape of Northern California has influenced many artists. Among writers, it has been most often the poets (Janet Lewis, Denise Levertov, John Haines, and Alistair Elliot, to cite a few examples) who have explicitly paid homage to that natural setting through their poetry. A line from "Landscape near Bear Gulch Road" by Janet Lewis, who at 90 was the oldest artist-in-residence, gives a feeling for the impact of that natural setting:

> One tone, one visible substance, a splendor
> Of multiplicity and self-abandon,
> Beneath that band of intense blue
> Lacking which, the hill is incomplete.

There have been so many writers in residence (cf. Table I) that I will mention only the very first, Joyce Carol Thomas, a black writer from Berkeley, who managed to complete an entire autobiographical novel while in residence. It was a great omen that the following year, she won the National Book Award in the juvenile category for her very first book.

The impact of the environment is, of course, most easily demonstrated and understood in the visual arts, and I shall end my account with commentary on three such artists. One of the youngest artists who has ever been accepted is Seyed Alavi, an Iranian environmental artist, trained in California and now living in Oakland. His works include "Limited Image," a cluster of silk tulips on one of our hillsides; "Tiger among Redwoods," featuring a life-sized fiberglass tiger in our forest; and "Echo," for me the most moving of them all, because it shows the word "echo" in etched glass floating in a pool at the foot of a waterfall, the spot where my daughter's ashes had been scattered.

Compared with Alavi, the other two artists, both sculptors, worked on very different time scales of permanence. The Italian Mauro Staccioli, who was proposed to our program's selection panel by none other than Giuliano Gori, eventually came three times: first to explore the land, next to make preliminary drawings and plans, and finally to construct an untitled work consisting of five huge, abstract geometric structures, covered in gray stucco, that interact physically and visually with the old, live oak trees to which they are connected. This installation, ranging over an area of a couple of acres, has been one of the most photographed installations on our site. The largest of these five objects collapsed during the big October 1989 earthquake. A year later, in response to that challenge, Staccioli repaired the damaged piece by doubling its size and then juxtaposing it with a 55-foot-tall semi-ellipse. Although Staccioli has numerous installations throughout the world, the largest one being a 150-foot-tall structure at the Olympic Games site in Seoul, only Gori's *Fattoria de Celle* and our site have offered him the opportunity to explore on a grand scale the marriage of urban stark materials with a forest's intrinsic, living complexity.

My last example of providing a sculptor with the opportunity of siting his work in unspoiled nature is David Nash, who tries, to quote his words, "to acknowledge the relationships of the forest and those who work it: using their materials and tools and calling on their experience." Nash became particularly well known to the American public through the exhibit "British Sculpture since 1965," which traveled through many of the major American museums, including the Chicago Art Institute and the Hirshhorn Museum in Washington, DC. The emotional impact of his work on me is best described in the following account of what happened on Thanksgiving Day, 1989, which was the coda to "A Scattering of Ashes" in my autobiography.

We had already been hiking for four hours in search of some felled redwood trunks of at least five-foot diameter. It's the minimum size David Nash requires for the three-part sculpture he intends to site around some burnt-out giant redwood stumps that can still be found, here or there, on our property from nineteenth-century logging practices. Nash is one of the most distinguished artists we have had in residence at our program. A British sculptor, now working in Wales, he first came to the Foundation in 1987 at the time of his retrospective exhibition at the San Francisco Museum of Modern Art. Although wood ("King of the Vegetables," he calls it) is his sole medium, and chain saw and ax are his principal tools, he had never before handled redwood or madrone, the two most prevalent species in our forest.

During his first stay, he had created a group of madrone sculptures for a highly successful show in Los Angeles; in addition, out of a huge redwood trunk that had lain for decades in Harrington Creek, Nash had fashioned "Sylvan Steps"—a Jacob's ladder rising at a steep angle out of the water into the sky. When he first selected that site—accessible only along the creek bed by clambering over rocks and fallen timber—he had had no inkling that only a few hundred feet upstream we had scattered in 1978 my daughter's ashes. Within a minute, some flecks of Pami's ashes must have floated past the spot where Nash's steps now rise into the air. "Sylvan Steps" is a magically simple sculpture that many subsequent artists have drawn, photographed, or written about.

But now, two years later during his second residency, we cannot find the massive log he needs. During our morning hike we have located four sites in the forest where blackened trunks rise out of the bracken—just the right backdrop for the scorched pyramid, cube, and ball Nash plans to shape, but the right arboreal progenitor for these forms is still missing. Of course, we cross the shadow of many a living redwood giant, but cutting one is out of the question. That's when I recall that some selective logging has just been completed on our neighbor's land across Bear Gulch Road; only a few days ago, I had followed impatiently a slow-moving truck stacked high with redwood logs. Perhaps "our" piece had not yet been removed.

I don't expect anyone working there on Thanksgiving Day, but after climbing over the locked gate and walking down the forest road, inches deep in dust (it had not rained for weeks), we hear in the distance the grinding of gears. Soon we come upon a mammoth tractor setting up erosion breaks to preserve the road bed during the winter rainy season.

"Have you moved out all the logs?" I shout up to the cabin after the bearded driver has shut off the thundering engine. "We need . . ." I say, and then explain who David Nash is, and why we are searching for a special fallen redwood rather than a turkey for Thanksgiving.

"All gone," he says, but then remembers. "A big one fell across the fence near the property line. Probably years ago . . . in some storm." According to him, it was partly rotten—sufficiently so that it had not been worth their while to haul it to the mill. Nash is dubious that it will do, but I say, "Let's look anyway."

We follow the man's directions to the fence a half mile down the logging road. When we finally come upon it, I am dumbfounded. Eleven years ago, I had hobbled here as fast as my stiff leg would carry me; but I had come from the opposite direction—down the meadow from our side of the property, toward this fence across which the massive trunk now lay, broken into three enormous pieces. It is the spot where my daughter had killed herself, where I have never dared to return. We find the rot to be only superficial; the wood is precisely what David Nash had been seeking all Thanksgiving long.

Acknowledgments

Acknowledgments

The chapters in this volume first appeared or were presented as follows. Grateful acknowledgment is made to the publishers for permission to reprint previously published material.

Chapter 1: From the Lab into the World

Carl Djerassi presented the Priestley Medal Address at the awards ceremony during the American Chemical Society's 203rd National Meeting in San Francisco in 1992. The address was originally published as "From the Lab into the World" in *Chemical & Engineering News* (April 6, 1992). Copyright 1992 by the American Chemical Society.

Chapter 2: Parentage of the Pill

This chapter is based on four separate texts. The *Bulletin of the American Academy of Arts and Sciences* is the original publisher of "Birth Control after 1984 Revisited" (Vol. 32, No. 1). Copyright 1978 by the American Academy of Arts and Sciences. "The Making of the Pill" first appeared in *Science 84* magazine (No. 5, p. 127). Copyright 1984 by the American Association for the Advancement of Science. "The Chemical History of the Pill" can be found on pages 339–361 of *Discoveries in Pharmacology 2,* edited by M. J. Parnham and J. Bruinvels. Copyright 1984 by Elsevier Science Publishers. "Steroid Research at Syntex: 'The Pill' and Cortisone" was originally published in *Steroids* (Vol. 57). Copyright 1992 by Butterworth-Heinemann. Excerpts from the Gregory Pincus Memorial Lecture (November 1993, Laurentian Hormone Conference) have not been published before.

Chapter 3: Progestins in Therapy: Historical Developments

"Progestins in Therapy: Historical Developments" was originally published as Chapter 1 of *Progestogens in Therapy,* which was edited by G.

Benagiano, P. Zulli, and E. Diczfalusy and published by Raven Press of New York. Copyright 1983 by Raven Press.

Chapter 4: The Manufacture of Steroidal Contraceptives: Technical Versus Political Aspects

"The Manufacture of Steroidal Contraceptives: Technical Versus Political Aspects" was originally published as part of the Proceedings of the Royal Society of London (B, Vol. 195, p. 175) in 1976. Copyright 1976 by the Royal Society.

Chapter 5: Birth Control after 1984

"Birth Control after 1984" is adapted from an article by the same name in *Science* magazine (Vol. 169, p. 941) that was published in 1970. Copyright 1970 by the American Association for the Advancement of Science.

Chapter 6: Reversible Fertility Control

"Reversible Fertility Control" was originally published in Volume III (p. 55) of the Proceedings of the XXIIIrd International Congress of Pure and Applied Chemistry. Copyright 1971 by Butterworths.

Chapter 7: Male Contraception

"Male Contraception" was originally published as "Future Prospects in Male Contraception" in *The Stanford Magazine* (Vol. 7, No. 1, pp. 58–63, and 71, Spring/Summer). Copyright 1979 by the Stanford Alumni Association.

Chapter 8: New Contraceptives: Utopian or Victorian

"New Contraceptives: Utopian or Victorian" was published in 1991 in *Science and Public Affairs* (London; Vol. 6, p. 5). The text of the article was based on an Evening Technology Lecture, given at the Royal Society in May 1991. Copyright 1991 by the Royal Society.

Chapter 9: Searching for Ideal Contraceptives

"Searching for Ideal Contraceptives" was originally published in the November/December 1985 issue of *Society*. Copyright 1985 by Transaction Publishers.

Chapter 10: Steroid Contraceptives in the People's Republic of China

"Steroid Contraceptives in the People's Republic of China" was originally published in the Sept 6, 1973, issue of the *New England Journal of Medicine* (Vol. 289, pp. 533–535). Copyright 1973 by the Massachusetts Medical Society.

Chapter 11: The Bitter Pill

Science magazine (Vol. 245, p. 356) originally published the article "The Bitter Pill" in 1989. That article was based on the Gustavus John Esselen Award for Chemistry in the Public Interest address at Harvard University on April 6, 1989, and is an expanded version of a talk presented at the annual meeting of the Institute of Medicine of the National Academy of Sciences on October 19, 1988. Copyright 1989 by the American Association for the Advancement of Science.

Chapter 12: Injectable Contraceptive Synthesis: An Example of International Cooperation

"Injectable Contraceptive Synthesis: An Example of International Cooperation" was coauthored with Pierre Crabbé and Egon Diczfalusy, and originally published in *Science* magazine (Vol. 209, p. 992) in 1980. Copyright 1980 by the American Association for the Advancement of Science.

Chapter 13: Future Methods of Fertility Regulation in Developing Countries: How to Make the Impossible Possible by December 31, 1999

"Future Methods of Fertility Regulation in Developing Countries: How to Make the Impossible Possible by December 31, 1999" was originally published in *Research on the Regulation of Human Fertility* (E. Diczfalusy and A. Diczfalusy, Eds.; pp. 235–249) by Scriptor in 1983. Copyright 1983 by Scriptor Publisher.

Chapter 14: A High Priority? Research Centers in Developing Nations

"A High Priority? Research Centers in Developing Nations" is based on text originally published in *Bulletin of the Atomic Scientists* in January

1968 (pp. 23–27). Copyright 1968 by the Educational Foundation for Nuclear Science. Material was also taken from *Views of Science, Technology and Development* (E. and V. Rabinowitch, Eds.), published by Pergamon Press (Oxford). Copyright 1975 by Pergamon Press.

Chapter 15: Pugwash, Population Problems, and Center of Excellence

"Pugwash, Population Problems, and Center of Excellence" has been adapted from three texts: "Research Centers of Excellence in Developing Countries" (pp. 159–162) and "A Possible Catalytic Role for Pugwash in Defining New Contraceptive Methods for Developing Countries" (pp. 163–164) are both from the Proceedings of the 20th Pugwash Conference on Science and World Affairs. Copyright 1970 by the Pugwash Conference. Material was also taken from "Pugwash Population Problems of Centers of Excellence," from the Proceedings of the 21st Pugwash Conference on Science and World Affairs (pp. 183–188). Copyright 1971 by the Pugwash Conference.

Chapter 16: Insect Control of the Future: Operational and Policy Aspects

"Insect Control of the Future: Operational and Policy Aspects" was originally published in *Science* (Vol. 186, p. 596) in 1974 and was coauthored with Christina Shih-Coleman and John Diekman. Copyright 1974 by the American Association for the Advancement of Science.

Chapter 17: Pesticide Development: Sociological and Etiological Background

"Pesticide Development: Sociological and Etiological Background" is taken from material originally published on pages 712–713 of the Proceedings of the XV International Congress of Entomology (Aug 27–29, 1976). Copyright 1977 by the Entomological Society of America.

Chapter 18: Planned Parenthood for Pets?

"Planned Parenthood for Pets?" was originally published in *Bulletin of the Atomic Scientists* in January 1973 and was coauthored with Andrew Israel and Wolfgang Jöchle. Copyright 1973 by the Educational Foundation for Nuclear Science.

Chapter 19: Research Impact Statements

"Research Impact Statements" was originally published as an editorial in the July 13, 1973, issue of *Science* magazine (Vol. 181, p. 115). Copyright 1973 by the American Association for the Advancement of Science.

Chapter 20: My Mom, the Professor

"My Mom, the Professor" was a letter to the editors published in *Science* magazine (Vol. 239, p. 10) in 1988. Copyright 1988 by the American Association for the Advancement of Science.

Chapter 21: Illuminating Scientific Facts through Fiction

"Illuminating Scientific Facts through Fiction" was originally published in the "Commentary" section of *The Scientist* (Vol. 4, p. 16). Copyright 1990 by The Scientist.

Chapter 22: Mentoring: A Cure for Science Bashing?

"Mentoring: A Cure for Science Bashing?" was originally published in the "Viewpoint" section of the Nov. 25, 1991, issue of *Chemical & Engineering News* on pages 30–33. Copyright 1991 by the American Chemical Society.

Chapter 23 Basic Research: The Gray Zone

"Basic Research: The Gray Zone" was originally published in *Science* magazine on August 20, 1993. Copyright 1993 by the American Association for the Advancement of Science.

Chapter 24: Some Forms of Art Patronage

"Some Forms of Art Patronage" was originally published in the *Bulletin of the American Academy of Arts and Sciences* (Vol. 44, p. 51) in 1991. Copyright 1991 by the American Academy of Arts and Sciences.

Index

Index

Copy editing: Janet S. Dodd and Elizabeth Wood
Indexing: Janet S. Dodd
Production: Donna Lucas
Acquisition: Robert N. Ubell
Text and cover design: Wilson Graphics & Design (Kenneth J. Wilson)
Cover photo: Michael Birt

Composition: Vincent L. Parker
Printed and bound by Maple Press, York, PA

Other ACS Books

Biotechnology and Materials Science: Chemistry for the Future
Edited by Mary L. Good
160 pp; clothbound, ISBN 0–8412–1472–7, paperback, ISBN 0–8412–1473–5

Chemical Demonstrations: A Sourcebook for Teachers
Volume 1, Second Edition by Lee R. Summerlin and James L. Ealy, Jr.
192 pp; spiral bound; ISBN 0–8412–1481–6
Volume 2, Second Edition by Lee R. Summerlin, Christie L. Borgford, and Julie B. Ealy
229 pp; spiral bound; ISBN 0–8412–1535–9

The Language of Biotechnology: A Dictionary of Terms
By John M. Walker and Michael Cox
ACS Professional Reference Book; 256 pp;
clothbound, ISBN 0–8412–1489–1; paperback, ISBN 0–8412–1490–5

Cancer: The Outlaw Cell, Second Edition
Edited by Richard E. LaFond
274 pp; clothbound, ISBN 0–8412–1419–0; paperback, ISBN 0–8412–1420–4

The ACS Style Guide: A Manual for Authors and Editors
Edited by Janet S. Dodd
264 pp; clothbound, ISBN 0–8412–0917–0; paperback, ISBN 0–8412–0943–X

Chemistry and Crime: From Sherlock Holmes to Today's Courtroom
Edited by Samuel M. Gerber
135 pp; clothbound, ISBN 0–8412–0784–4; paperback, ISBN 0–8412–0785–2

Steroids Made It Possible
By Carl Djerassi
206 pp; clothbound, ISBN 0–8412–1773–4

Writing the Laboratory Notebook
By Howard M. Kanare
145 pp; clothbound, ISBN 0–8412–0906–5; paperback ISBN 0–8412–0933–2

Understanding Chemical Patents: A Guide for the Inventor
By John T. Maynard and Howard M. Peters
183 pp; clothbound, ISBN 0–8412–1997–4; paperback ISBN 0–8412–1998–2